Y0-BTC-450

CSCPRC REPORT NO. 7

Astronomy in China

A Trip Report of the American Astronomy Delegation

Edited by Leo Goldberg and Lois Edwards

Submitted to the Committee on Scholarly Communication
with the People's Republic of China

NATIONAL ACADEMY OF SCIENCES
Washington, D.C.   1979

NOTICE: The exchange visit of the Astronomy Delegation to the
People's Republic of China was supported by a grant from the National
Science Foundation. This visit was part of the exchange program oper-
ated by the Committee on Scholarly Communication with the People's
Republic of China, founded jointly in 1966 by the American Council of
Learned Societies, the National Academy of Sciences, and the Social
Science Research Council. Sources of funding for the Committee include
the National Science Foundation, the Department of State, and the Ford
Foundation.

The Committee represents American scholars in the natural, medical,
and social sciences, as well as the humanities. It advises individuals
and institutions on means of communicating with their Chinese colleagues, on China's international scholarly activities, and on the
state of China's scientific and scholarly pursuits. Members of the
Committee are scholars from a broad range of fields, including China
studies.

Administrative offices of the Committee are located at the National
Academy of Sciences, Washington, D.C.

The views expressed in this report are those of the members of the
Astronomy Delegation and are in no way the official views of the Committee on Scholarly Communication with the People's Republic of China
or its sponsoring organizations--the American Council of Learned Societies, the National Academy of Sciences, and the Social Science Research Council.

International Standard Book Number 0-309-02867-1

Library of Congress Catalog Card Number 78-78137

Available from:

Office of Publications
National Academy of Sciences
2101 Constitution Avenue, N.W.
Washington, D.C. 20418

Printed in the United States of America

CONTRIBUTORS

VICTOR M. BLANCO, Director, Cerro Tololo Inter-American Observatory,
La Serena, Chile

E. MARGARET BURBIDGE, Professor of Physics, University of California,
La Jolla, California

LEO GOLDBERG, *Chairman of the Delegation,* Kitt Peak National Observatory, Tucson, Arizona

DAVID S. HEESCHEN, Director, National Radio Astronomy Observatory,
Charlottesville, Virginia

GEORGE H. HERBIG, Professor of Astronomy, Lick Observatory, University
of California, Santa Cruz, California

ALLAN R. SANDAGE, Astronomer, Hale Observatories, Pasadena, California

MARTIN SCHWARZSCHILD, Higgins Professor of Astronomy, Princeton
University Observatory, Princeton, New Jersey

NATHAN SIVIN, Professor of Chinese Culture and of the History of
Science, Department of Oriental Studies, University of Pennsylvania,
Philadelphia, Pennsylvania

HARLAN J. SMITH, Director, McDonald Observatory, University of Texas,
Austin, Texas

CHARLES H. TOWNES, University Professor of Physics, University of
California, Berkeley, California

PREFACE

This is a report of a 25-day visit to the People's Republic of China
by an American delegation of nine astronomers and one historian of
Chinese science. The delegation arrived in Peking by air from Tokyo
on September 29, 1977, and departed Canton by train for Hong Kong on
October 24, 1977. The visit was arranged jointly by the Committee on
Scholarly Communication with the People's Republic of China (CSCPRC)--
which is sponsored by the National Academy of Sciences, the Social
Science Research Council, and the American Council of Learned Societies--
and the Scientific and Technical Association of the People's Republic
of China (STAPRC) as one in the series of scholarly exchanges sponsored
by the two organizations.

During their 25-day stay in China, the American astronomers traveled
more than 6,000 km, while visiting the cities and surroundings of
Peking, Shanghai, Nanking, Kunming, Kweilin, and Canton. The delegation was guided on its tour by Li Ming-te, staff member of the STAPRC.
His organizational skill, tact, unfailing good humor, and superb command of English were invaluable assets to the delegation. He was
assisted by Wu Ling-an, a physicist of the Institute of Physics of the
Chinese Academy of Sciences (CAS) in Peking, whose many kindnesses and
service as the scientific interpreter and guide were very much appre-
ciated by the delegation. Mr. Teng Ting-yu, also on the staff of the
STAPRC, handled all of the many logistical problems with great smooth-
ness and efficiency. Mr. Richard Bock of the U.S. Liaison Office in
Peking also traveled with the delegation and contributed greatly by
his knowledge of the Chinese language and of the customs and politics
of the country. (The delegation and traveling party with hosts at Purple
Mountain Observatory appear in Figure 1.) The delegation was also for-
tunate to have the company of Mr. Alex DeAngelis, of the CSCPRC staff,
from the time of its arrival in Tokyo until its departure from Peking.
His experience in escorting other delegations in China, as well as his
mastery of the language, helped enormously in cushioning the shock of
sudden exposure to a totally unfamiliar environment and in accelerating
the process of acclimatization.

The report that follows represents the combined efforts of all
members of the delegation. Prior to entering China, the delegation
met in Tokyo on September 28 and agreed upon a provisional list of
topics that would constitute its report. Each member was assigned

v

FIGURE 1   The astronomy delegation and traveling party with hosts at
Purple Mountain Observatory. First row: Wu Ling-an, Interpreter,
Institute of Physics, CAS; Teng Ting-yu, Staff, STAPRC; Kung Shu-mo,
Purple Mountain Observatory (PMO); Hsiang Te-lin, PMO; Margaret
Burbidge; Martin Schwarzschild; Tseng Wu-chu, PMO Secretariat; Victor M.
Blanco. Second row: Li Ming-te, Staff, STAPRC; Ts'ui Lien-shu, Nanking
University; Chao Wen-piao, PMO Administrator; George Herbig; Leo
Goldberg; Harlan J. Smith; Allan Sandage; David S. Heeschen. Third
row: Chang Yu-che, Director, PMO; Nathan Sivin; Richard Bock, U.S.
Liaison Office in Peking; Charles H. Townes.

primary responsibility for one or more topics with the understanding
that all would be welcome to contribute to any and all parts of the
report. The topical outline was revised from time to time as the
visit progressed and again immediately after the delegation's return
to the United States. In January 1978, the delegation met in
Washington, D.C., to review first drafts of the various sections,
most of which had been previously circulated and revised, and to decide
upon a final list of chapter headings and writing assignments. In
view of the multiple input to the report, the delegation decided not
to attribute individual authorship to its various sections. Editorial
work began in March and was completed by the end of July 1978. We

acknowledge with thanks the valuable editorial contributions of Mr. Alex DeAngelis and Ms. Lois Edwards of the CSCPRC staff.

LEO GOLDBERG
Chairman of the Delegation

CONTENTS

1 INTRODUCTION   1

2 HISTORY OF ASTRONOMY   7

3 ORGANIZATION AND ADMINISTRATION   22

4 INSTITUTIONS FOR RESEARCH AND EDUCATION   25

5 RESEARCH INSTRUMENTS AND PROGRAMS   30

  Solar Physics, 30
  Solar System, 35
  Time Service and Related Astronomy, 36
  Satellite Observations and Orbital Determinations, 39
  Stellar Astrophysics, 42
  Radio Astronomy, 45
  Extragalactic Astronomy, 47
  Electronic Computers, 53
  History of Astronomy, 55

6 INSTRUMENTS AND FACILITIES UNDER CONSTRUCTION   58

7 EDUCATION AND RESEARCH IN ASTRONOMY AT UNIVERSITIES   64

8 DISCUSSION AND ISSUES   68

  Interaction Between Chinese Astronomers and the
    International Community, 68
  Options for Future Development of Optical Instruments, 70
  Options for Future Development of Radio Astronomy Instruments, 71

Theoretical Astrophysics, 72
Publication Problems, 73

APPENDICES

A Principal Instruments, Research Programs, and Staff of the
   Various Institutes and University Departments   75
B Summary Itinerary   92
C Lectures   99
   Lectures Presented by the Chinese, 99
   Lectures Presented by the Delegation, 103
D Comments on the Selection of an Observing Site for Optical
   Astronomy   106

1

INTRODUCTION

Prior to 1950, relations between American and Chinese astronomers
had been close. A number of Chinese astronomers had studied and
worked at American observatories and later facilitated exchanges of
information between the two countries. Beginning with the Korean War,
direct communication ceased altogether and after 1960, when the
People's Republic of China withdrew from the International Astronomical Union, contact between American and Chinese astronomers
virtually came to an end. For a time, it was still possible to keep
in touch with astronomical research in China by reading *Acta Astronomica Sinica,* but even that source of information was lost in 1966,
when the publication of research results was suspended, not resuming
until 1974.

American astronomers had been deeply disappointed and frustrated
by the prolonged involuntary separation from their Chinese colleagues. Beginning in 1973, reports from visits to China by individual astronomers from the United States and other countries had been
quite encouraging in indicating a high level of astronomical activity
in China. However, these reports only sharpened our desire to broaden
and deepen our contacts, and therefore we were delighted when a delegation of Chinese astronomers spent 1 month visiting observatories in
the United States in November-December 1976. This visit was quickly
followed by an invitation from China for a return visit by an
American delegation.

In view of our very limited knowledge of Chinese astronomy, we
felt that our mission should be to survey astronomy in China by visiting major observatories, institutes, and universities; to establish
friendly relationships with Chinese astronomers and administrators;
and to lay the groundwork for future exchanges. From published reports
by other visitors and from the 1976 visit of the Chinese astronomers,
we knew that the Chinese were very active in a number of subfields,
especially celestial mechanics, astrometry, time determination, solar
and stellar physics, and radio astronomy. We sought to become acquainted with Chinese goals for the near-term development of these
and other subfields of astronomy and to familiarize ourselves not
only with current and projected research, but also with the administrative framework in which it is conducted, the levels of financial

and technological support, and the educational system in which astronomers are trained.

To achieve our purposes, we requested an itinerary that included visits to Peking Observatory and three of its observing stations, Peking University, Purple Mountain Observatory, Nanking University, the Nanking Astronomical Instruments Factory, Shanghai Observatory, and Yünnan Observatory in Kunming. All of our requests were granted (see Appendix B). In addition, our stay in China was enlivened and enriched by a variety of excursions to places of exceptional historical and cultural interest and extraordinary beauty. We were also priviledged to participate in the celebration of the twenty-eighth anniversary of the founding of the People's Republic of China in Peking.

We could hardly have arrived in China at a more propitious time, just at the beginning of the celebration of the National Liberation Day, and shortly after the publication of a circular by the Central Committee of the Communist Party, which called for the holding of a national conference in science. In the words of Chou P'ei-yüan, the circular was "a mobilization order for scaling the heights in science and technology. We shall make an all-out effort to catch up with and overtake advanced world levels and modernize our industry, agriculture, national defense, and science before the end of this century. This goal must be attained." We had the feeling of being present at a very important juncture in the long and fascinating history of China and that in years to come we would look back upon our visit as a time of great historical significance.

We were given the impression that astronomy will be an important part of the advancing front of science and technology in China, and indeed we recalled the words of Chairman Mao, who once said that every great country should be engaged in seeking to discover how the universe began and how it evolved. Our brief encounter with Chinese astronomy has convinced us that Chinese astronomers have the talent and the motivation to be in the forefront of that great search and that they fully understand and can deal with the present obstacles in their path. Chief among them are the needs to add rigor to the selection of students, to improve the quality of their training, and to provide experimenters with equipment suitable for research on the frontiers of astronomy. We were greatly impressed by the brilliance of much of the theoretical work we saw, although we noted that it suffered considerably from lack of contact with advanced observational research. We feel that relatively long visits to one or more American observatories would be of enormous benefit both to the Chinese theorists and to the observers who would profit from working with the most advanced equipment. Many of the young people we met would need little preparation, other than in English language study, in order to benefit to the full from such visits.

Our hosts at the various institutions had prepared a full and lively program providing for exchanges of information between our delegation and the Chinese astronomers. These exchanges took the form of formal lecture sessions in which both sides participated (see Appendix C) and of formal discussion meetings to which we and the Chinese came with prepared lists of questions.

The only member of the delegation who was not a professional astronomer was Nathan Sivin, a historian who has published on every field and
every period of science in traditional China, as well as on the encounter of Chinese and European science in early modern times. Dr. Sivin
accepted the CSCPRC's invitation to join the delegation with the understanding that his itinerary would be the same as that of the other
members, and he was prepared to spend his entire time in China visiting
observatories, interviewing astronomers, and contributing to the activities of the delegation generally through his knowledge of China's
language, culture, and history. He did take part in most of the collective activities of the delegation, but in addition an opportunity
unexpectedly arose to contribute in another way to the aims of the
exchange program.

When the delegation visited Sha-ho Observing Station outside Peking
on September 30, among the Chinese welcoming group was Po Shu-jen, a
historian of astronomy who had published several important essays in
the early 1960's. The day after this meeting, Sivin was invited to
attend a series of meetings with a large group of historians of science,
mostly of astronomy, in Peking. In Shanghai and Nanking, Sivin was
invited to similar separate meetings with groups of historians who
were working at the observatories of the two cities, as well as at
Shanghai Normal University and in the Planning Group for the Shanghai
Planetarium. In Kunming two reports by astronomers on historical
research were delivered to the whole delegation.

So far as the CSCPRC is able to ascertain, this is the first time
since the Cultural Revolution that a humanist member of an American
delegation has been able to hold extended working discussions with
groups of Chinese colleagues. This was due in part, no doubt, to the
change in political climate that was perceptible in many other ways
at the time the delegation arrived. The fact that the invitation came
unexpectedly and only after Sivin's meeting with Po Shu-jen suggests
that the Chinese historians arranged the meetings after it had been
ascertained that they could be carried out entirely in Chinese (which
meant not only that greater informality was possible, but also that
about three times as much ground could be covered when no interpretation was necessary and with less risk of misunderstanding).

The meetings with historians of science have opened channels of
communication that have been closed for more than a decade, as research in the humanities and social sciences were more strongly
affected by the Cultural Revolution than was scientific work.

The organized lectures and discussions were only one of the means
by which information was gathered for this report. Equally important
were the conversations held with our hosts during dinners, in hotel
rooms, during tours of observatories and universities, and on sight-
seeing excursions. One particularly memorable occasion was a rainy
evening at Hsing-lung Station of Peking Observatory, where four members
of the delegation spent a rainy night and carried on a free and un-
inhibited discussion with about 20 Chinese astronomers during dinner
and in the evening until well after midnight. The Chinese were full
of questions about the design of telescopes and auxiliary instrumentation, the measurement and analysis of spectra of quasi-stellar

objects (QSO's) and galaxies, etc. On another evening in Nanking, we entertained a group of nine engineers of the Nanking Astronomical In- struments Factory at our hotel, where the main subjects for discussion were the design and construction of optical telescopes and accessories. On still another occasion at the hotel, members of the Department of Astronomy of Nanking University compared notes with several of our group on the selection and training of astronomers.

It was a particular honor and pleasure for our delegation to spend a fair number of hours in the company of two of China's most distin- guished physicists and senior statesmen, Chou P'ei-yüan, acting director of the Chinese Scientific and Technical Association, president of Peking University, and vice-president of the Chinese Academy of Sciences, and Wu Yu-hsun, vice president of the Chinese Academy of Sciences. Chou hosted the welcoming banquet for our delegation at the Peking Roast Duck Restaurant and was in turn the guest of honor at the return ban- quet that we gave at the Szechuan Restaurant, also in Peking. In the late 1920's and early 1930's, Chou studied cosmology in Pasadena at Cal Tech and at Princeton and is still active in research on the theory of turbulence. Wu studied at Princeton with K. T. Compton, and indeed the phenomenon by which energy is transferred during the scattering of X-ray photons by electrons is known in China as the Compton-Wu effect. Professor Wu was our host at the splendid banquet given by Chairman Hua in the Great Hall of the People on the evening of September 30 to initiate the celebration of the National Liberation Days.[*] Both men were most eloquent in describing their hopes for the future of science and technology in China while showing themselves to be clearly aware of the many diffi- cult problems that must be solved if the PRC is to catch up with the most advanced countries in science and technology by the year 2000.

Our Chinese hosts were extremely considerate in maximizing the opportunities for communicating with our delegation. When going be- tween cities, we usually traveled by air, except between Nanking and Shanghai, a comfortable $4\frac{1}{2}$-hour train ride through rich farm country. On trips to observing stations and excursions to places of esthetic and cultural interest, however, we always rode in a convoy of half a dozen gray "Shanghai" sedans. Apart from the lead vehicle, whose function was to open a path through the sometimes dense array of com- mercial vehicles and bicycles, each car usually contained a leading Chinese astronomer or administrator, an interpreter, who might also be an astronomer, and two Americans. Since the duration of each drive varied up to more than 4 hours, there was often ample time for ex- change of ideas and in-depth exploration of the scientific and political scene in both the United States and the People's Republic of China. From the many anecdotes reported by members of the delegation, we have chosen a few to illustrate the range of the subjects discussed and the deep interest of the Chinese in keeping in touch with the latest developments in astronomy in the United States.

*We learned with sorrow of Professor Wu's death, shortly after our return to the United States.

During our visit to Nanking, we were asked by a member of the Purple
Mountain Observatory staff about the U.S. Space Telescope: its size,
the instruments to be carried in it, and the plans for extragalactic
research with it. We were also asked about existing telescopes on
Mauna Kea and plans for the future, and the prospects for construction
of a very large ground-based telescope. We were told that while the
Chinese government had no space astronomy program at that time, astronomers recognized the importance of such a program and hoped to be involved in one sometime in the future. At the first national science
conference since 1950, held in Peking from March 18 to 31, 1978, it was
announced that space science and technology would be one of eight sec-
tors to receive priority in the allocation of resources for research
during the coming 8 years. The Chinese made it clear that they had no
definite plans to build large optical telescopes beyond the 2-m telescope projected for Peking Observatory, but at the same time their
questions indicated that they were considering various possibilities.
For example, at Yünnan Observatory, which is slated to become the major
center for astronomy in the south of China, Director Wu Min-jan spoke
of a 4-m telescope as "their dream" for the future, and in the course
of an informal evening meeting with Goldberg, Schwarzschild, and Townes,
Hung Ssu-yi, secretary of Peking Observatory, and Wang Shou-kuan, the
leading radio astronomer, asked whether in our opinion China should
attempt a very large next-generation telescope (NGT) after the completion of their 2-m, or whether there should be a 4-m telescope as an
intermediate step. The meeting was a wide-ranging discussion of such
topics as the 10-year gap in the training of scientists as a result of
the closing and alteration of the universities that came with the
Cultural Revolution, the staffing of observatories in part by inade-
quately trained young people, which was attributed to the "Gang of
Four," mechanisms for interaction between theorists and experimental-
ists, the expansion of exchanges between China and the United States,
the resumption of graduate work in the universities, selection pro-
cesses for employees at the observatory, and the system of dual management of institutes by scientific directors under the leadership of
party committees.

The Chinese astronomers are looking forward to the completion of
their 2-m optical telescope, which will be the largest such instrument
in China for a number of years and will therefore play a major role in
the near-term development of astronomy in the country. For this
reason, they seek information on the organization of national and
university observatories in the United States. During the car ride
from Kunming to the Stone Forest, one of us was asked many questions
on the operation of national and university facilities and, in particular, questions on the ways in which observing time is allocated. The
Chinese astronomers are well aware that some observatories in the
United States are run by one university, others by one or more universities or institutions, and still others as national facilities primarily for the benefit of the entire U.S. astronomical community.

The Chinese seem very concerned about how long it will take for
China to catch up with more advanced countries in highly technical
fields, such as infrared astronomy, in which there seems to be no

activity in China at the present time. Chou Tsun-po, director of Shanghai Observatory, who is not a scientist, asked many pertinent questions about Kitt Peak National Observatory, its organization and administration, especially about the special requirements of infrared astronomy, including such details as the temperatures of liquid nitrogen and helium. Liquid nitrogen is apparently commercially available in China and easy to obtain. Liquid helium is manufactured at the Institute of Physics in Peking at rates of up to 25 ℓ per hour.

In all of these encounters, both formal and informal, we found the Chinese fully responsive and forthcoming in answering questions and obviously eager for opinions and advice on the work they were doing. Moreover, the cordiality and enthusiasm with which we were received made it very easy for us to establish a friendly rapport with the Chinese astronomers and administrators after only 1 or 2 days of association, which helped enormously to speed the two-way flow of information.

2

HISTORY OF ASTRONOMY

The development of science policy since 1949, to the point where its results were observed by the Astronomy Delegation in 1977, has often been traced and has been capably summarized by our predecessors.[1]

Research in the physical sciences on the whole has not been concentrated in the universities but in research institutes under the central administration of the Chinese Academy of Sciences (CAS). Thus, although a certain proportion of teaching staff in the universities spend a portion of their time in research, most publishable research in astronomy is done in institutions designed for that single purpose. These institutions, like all others in the People's Republic of China, are expected to be fully responsive to developing and changing national goals.

Scientists everywhere over the past century or so have tended to think of themselves as citizens of an intellectual republic dedicated to an enhanced theoretical understanding of nature, which may be applied to the benefit of mankind. Although in China, as elsewhere, there is a consensus about the importance of the benefits, since 1949 there have been fateful and persistent differences of opinion in high places about whether these benefits are richest when scientists are following their own bent or when directions for research are prescribed by political authority. In a society that has given unprecedented status to the image of the educated and politically active peasant or worker, and has correspondingly devalued that of the highly trained professional specialist, the preference of a generation of scientists who were educated abroad for an autonomous and apolitical style of work has been unacceptable. One can view the Great Leap Forward and the Great Proletarian Cultural Revolution as extended phases of an experiment to weave science and technology firmly into the fabric of national life--much more firmly than most other societies expect them to be integrated--without cutting them away from their vital sources. The Chinese long ago rejected the notions that some worthwhile research is so basic that it can be pursued with no concern whatever for its consequences, and that the scientific quest must be so single-minded that scientists are entitled to a special exemption from mass-mobilization politics. This distinctive Chinese outlook was epitomized at one of the delegation's formal dinners by a former high military officer, now responsible at the provincial level for science and education. When asked how, in view of

the nature of science, there could be a characteristic Chinese or
Marxist form, he responded that, while the truths of science are universal and must be respected, the uses of science are political and
must be controlled. Certain vital questions remain open, we feel, above
all what emphasis should be put on research that offers immediate results
in industrial or agricultural production, as distinct from research
that can be justified only by its applications in the long run.

The changes of the last 20 years in policy toward scientific research and education have been greater in magnitude than the innovations of any of the other great nations. It would be nonsense to
argue that they represent a series of carefully planned experiments
that remained from first to last under the control of policymakers.
On the other hand, the effects of the last 20 years' developments on
science, as on education and economy, have provided data for an ongoing
assessment that our Chinese hosts believe is taking them closer to a
proper balance--a balance of expertise and "redness"--that will recon-
cile the continuing predominance of Maoist activism with the impera-
tives of survival in a technologically ever more complex world.

MODERN ASTRONOMY AND ITS CHINESE TRADITION

Astronomy was one of the major sciences of ancient China, along with
medicine, alchemy, and other enterprises even less familiar to Western
readers. Traditional medicine, at least medicine of a kind manifestly
derived from that practiced before the influence of the West became
important, is still practiced today. The same cannot be said of
astronomy.

Chinese astronomers before modern times, like their counterparts in
Europe, were concerned primarily with predicting the positions and
major phenomena of the Sun, Moon, and planets, as well as with observing changing features of the sky that the naked eye is able to reveal.
Today the problem of computing the ephemerides for public use has become too trivial to be of any further theoretical interest. The
telescope has enormously extended the depth to which astronomers
can search for changes in the visible universe. Observations are
valued to the extent that they throw light on questions about which the
ancient stargazers could only speculate, such as the origin, matura-
tion, and death of stars, and the forces that determine orbits in the
solar system. The very best ephemerides computed before the advent
of the telescope in Europe or China would certainly have been adequate
to any practical application then conceivable; but they would seem
crude today. The calculation of planetary motions, times of eclipses,
and so on, which occupied the entire careers of China's greatest
astronomers, are today a minor and quickly traversed episode in the
training of an undergraduate astronomy major. If one wishes to prepare for the study of pre-Newtonian astronomy, today's textbooks are a
great deal less servicable for learning the computational techniques
needed by a historian than those published two centuries ago.

Because modern astronomy has solved the classical problems and has
gone off in new directions, the relevance of its past is by no means

obvious. Most working astronomers in the United States, Europe, and Japan feel no need to look back at the concerns of their pre-modern predecessors. In China the native astronomical tradition is a matter of awareness among not only astronomers, but also the general public. The greatest astronomers of the last 2,000 years have been depicted on postage stamps (Figure 2), their innovations are displayed in museums, and their achievements are chronicled in children's books and popular magazine articles. This interest would not be remarkable if the antiquarian tendency of the old Chinese elite were still dominant, but to

FIGURE 2 Postage stamps depicting Chinese scientists and inventors and their achievements.

the contrary, the past is valued only to the extent that it can play a role in defining a socialist future. What is the reason for this apparent paradox?

The paradoxical quality of this abiding interest in the past has become even more apparent in the last decade. Soon after the Cultural Revolution began, the Research Institute for the History of Science was among several institutes in the Chinese Academy of Sciences that closed their doors until 1976. During this long period much of the normal work of the established academic disciplines in the humanities and social sciences was set aside for a protracted debate on what the national role of those disciplines should be. This does not mean that contributions to scholarship ceased during this period. To the contrary, the results of important new research on early astronomy, published again since 1973, have continued to enrich and reorient our understanding.

In order to understand the power that the astronomical past retains in contemporary China, it will be worth while to consider the character of that past, its interaction with foreign astronomical traditions, and the confluence of Chinese and Western lines of development in this century to form modern astronomical institutions.

THE CHARACTER OF CHINESE ASTRONOMY

A Chinese visiting Europe in 1400 would have found it technologically backward. This conclusion has been well documented over the last couple of decades by historians in China and elsewhere, and its pieces have been tied together and presented to a wide Western audience in Joseph Needham's *Science and Civilisation in China*. The superiority of Chinese material culture is not surprising, since by 1400 the West was still recovering from the near-collapse of a millennium earlier. By contrast, the practically uninterrupted growth of civilization made it possible for China to give more than it took in the international concourse of technology.

It is dangerous to leap to the conclusion that the situation of science equalled that of technology. Today the two kinds of activity are hardly divisible; mechanisms and devices define the changing frontiers of scientific exploration, and theoretical discovery generates technical innovation. Before the Scientific Revolution of the seventeenth century, on the other hand, science was largely a search for understanding, carried on by the educated minority, and had practically no effect on the techniques of craftsmen and builders. The latter passed down their skills privately to their children or apprentices. Scientists learned from the aspects of artisans' work that happened to come to their attention, but artisans had little access to science.

Attempts in Europe and the United States over the past century to prove that Chinese were incapable in science, and, more recently, that they were far ahead of the Occident in science until the time of Galileo and Newton, have tended to compare situations that had little in common.

It is true that great bronze armillary spheres and water-driven astronomical clocks were being built in China in the eleventh century, at a time when in Europe instrumentation was rudimentary and only a handful of people had the command of mathematics needed for the simplest astronomical computations. But this low level in Europe was a regression from the finely adjusted predictive models that Ptolemy had worked out by about A.D. 150. In the Middle Ages the transition from human power to the harnessed power of animals, wind, and water took many centuries; but once Ptolemaic astronomy returned to Europe from Islam it was not long until European scientists had mastered it and were moving on. Classical astronomy had similarly provided a stimulus earlier in Islam and India. As the next section will show, in astronomy, unlike technology and medicine, the net flow over the last thousand years, so far as we can trace it, was from India, Islam, and Europe (all of which used modified Ptolemaic traditions) toward China. What was the case earlier is still being warmly debated, but the substantial originality of Chinese astronomy can no longer be doubted. To demand that we demonstrate the superiority of one tradition over the other in order to find it worthy of study is a symptom of atrophied curiosity. It is a fair generalization that the astronomical traditions of East and West are comparable. Understanding both prepares us to recognize what was universal, and what was merely local variation, in each.

In both Chinese and European traditions, astronomical computations lay at the basis of the calendar, but they were a routine matter except in periods of calendar reform. The outcome of each calendar reform was a set of procedures that someone with no astronomical skill could follow to work out the annual calendar (or ephemeris) a simple step at a time.

In Europe the main employment of astronomers was in astrology. Chinese astronomical officials were supported for the same purpose. The function of astrology was, however, different in East and West.

The Hellenistic world evolved the horoscope, which made it possible to analyze the presumed influences that configurations of stars exerted upon individuals at the time of conception or birth. An astrologer was thus prepared to calculate the geometric relations of the Sun, Moon, and planets at any moment for anyone. Methods popular throughout Chinese society for casting individual fates were not entirely dissimilar, but they did not depend on the actual configuration of the sky. Astrology, in which unusual celestial events were observed and interpreted as omens, was done in the imperial court on behalf of the monarch. Its relation to observation and computation was quite different than in Europe.

According to the Chinese theory of monarchy, the ruler was fit to rule because of the accord he maintained with the cosmic order through his personal virtue and because he correctly performed certain indispensable rituals. This accord, at the same time charismatic and profoundly natural, enabled him to maintain an analogous order in the political realm. The state was, in other words, a microcosm. If the emperor lacked virtue or was careless in his duties, disorderly phenomena would appear in the sky (or in nature generally, since the macrocosm included the whole natural world) as a portent of disorder in the political sphere.

This theory divided celestial phenomena into those that were regular and could be computed and those that were irregular and unpredictable, and thus omens. Astronomers had two complementary tasks: to incorporate as many phenomena as possible in a correct calendar--actually an ephemeris that included, in addition to days and months, planetary phenomena and eclipses--and to observe unpredictable phenomena so that they could be interpreted and the emperor warned as early as possible that all was not well in his realm. The calendar was thus part of the ritual paraphernalia that demonstrated the emperor's legitimacy. Knowledge of astronomy could easily be manipulated and would be dangerous in the hands of someone trying to undermine the current dynasty. It was therefore generally believed that the proper place to practice astronomy was in the imperial court, and in certain periods it was illegal to do it elsewhere.

In ancient Babylonia computation was based on repetitive numerical procedures not unlike those that might be used in a simple computer program today. As cyclical periods were recognized, they could be imposed upon each other to approximate the complexities visible in the sky. For example, once the average apparent period of the Moon's travel through the stars was recognized with precision, study of variations in travel revealed that they had their own period, the anomalistic month; attempts much later to combine these two periods for more perfect prediction uncovered more subtle variations with their own periods.

The Greeks substituted geometrical models for the Babylonian algebraic approach. They were thus prepared to deal in a relatively direct and simple way with such problems as solar eclipse prediction, in which, no matter what the method, one must ultimately determine the intersection of the Moon's shadow cone with the three-dimensional surface of the earth.

The Chinese began with an approach grossly similar to that of the Babylonians. By the second century A.D. they were making assumptions about spatial relations as they designed their mathematical models, and before the end of the eleventh century they were experimenting with what might be called proto-trigonometry. Their ability to use what amounts to higher-order equations and the precision with which they determined the periods of celestial motions grew steadily until the Chinese calendar reached its highest point of sophistication shortly before 1300. Lunar eclipses could be predicted with sufficient accuracy by 100 B.C., so that the Chinese no longer reported them as omens. But solar eclipse techniques never attained the accuracy that Ptolemy had reached. Because what was unpredictable remained ominous, the imperial court tended to look abroad for solutions to the abiding problem of solar eclipses. This tendency was fateful for the development of astronomy within China and for the character of foreign influence.

Beneath the steady advance of computational astronomy in China lay a foundation not only of observational instruments, but also of data-recording systems and of institutions able to maintain records and standards for centuries on end. Joseph Needham has summarized what is characteristically Chinese in this foundation:[2]

1. The elaboration of large and complex astronomical instruments, both observational and demonstrational. Armillary instruments probably go back to the second century B.C., and Hsü Chen-t'ao has argued from recently discovered planetary records that they may have been invented as early as the fifth century B.C.[3] Instruments of this sort continued to be made until they were replaced by telescopes from Europe. The last major group of Chinese instruments, in fact, was cast in bronze at the initiative and under the direction of Jesuit missionaries in the seventeenth century, after the telescope had reached China. They may be seen in Peking today atop one of the last surviving sections of its city wall. Three fifteenth-century instruments of entirely Chinese type are preserved at Purple Mountain Observatory in Nanking (Figure 3).

2. The invention of the clock drive, and perhaps of the clock escapement itself, in a long series of great astronomical clocks in which a water-driven mechanism rotated that forerunner of the telescope, the sighting tube. Even in a device of this kind, about which we are best informed, built late in the eleventh century, the motion of the equatorial ring that carried the sighting tube on the armillary sphere would not have been smooth enough to give the instrument any value for observation. Its moving celestial globe and time indicators made it a splendid showpiece, another monument to the imperial charisma. The direct influence of Chinese mechanisms on the development of clocks and clock drives in Europe has not been traced.[4]

FIGURE 3 Abridged Armilla of 1437 at Purple Mountain Observatory, Nanking.

3. The maintenance of accurate dated records of such phenomena as eclipses, novas, comets, and sunspots over a longer continuous period than in any other civilization.[5]

4. The earliest known star catalogues embodying quantitative positional data (fourth century B.C., although what remains of these writings is apparently not earlier than about 70 B.C.). No less important historically, although not as old as analogous Mesopotamian documents in clay, are the ephemerides for the year 134 B.C., discovered in 1972, and the records of Jupiter, Saturn, and Venus for the period 244 to 177 B.C., excavated in 1973. Found in the same tomb as the latter was another manuscript, which portrays many types of comets and other astrological portents, the most ancient of its kind extant from any civilization.

5. A system of coordinates mainly oriented on the equator and the equatorial pole, like the one that in the modern West has replaced the classical celestial longitudes and latitudes. In Europe phenomena were recorded according to their location in 1 of 12 equal houses of the zodiac. This practice made it easy to measure angular distance from the nearest of 12 standard stars, and even to estimate longitude with the naked eye. In China the same purpose was served by 28 unequal divisions, the "lunar lodges," the determinative stars of which lay near the equator but were keyed to bright stars of the same right ascension in the vicinity of the celestial pole. Thus right ascension could be read even when the determinative star was below the horizon.

6. Among several early conceptions of the universe, one in which it was boundless and in which the stars floated in empty space. Despite (or perhaps because of) its audacity, this conception had little influence in cosmological debates, and none that has been documented on the practical work of astronomers.

The third of these characteristics deserves particular stress. Despite the limitations of pretelescopic observation, ancient records are useful to modern astronomers in many ways. The orbital periods of such celestial bodies as Halley's comet and the frequency of sunspot cycles have been determined with confidence; detailed descriptions of supernova explosions that coincide with today's radio sources have been compiled.[6] The most rigorous research of these sorts has been centered on the Chinese records, supplemented primarily by those of Japan, Korea, and Islam. European records are of only limited use for most of these purposes, since those who knew little astronomy seldom left elaborate records of phenomena, and the churchmen and educated laymen who knew a good deal of astronomy before the seventeenth century accepted the Aristotelian dictum that there could be no change in the skies beyond the sphere of the Moon. Only someone ignorant of epistemology would argue that the learned astronomers of the ancient West were ignoring sunspots and stellar explosions; it would be more accurate to say that they did not see them. In the Far East the astrological tradition provided bureaucrats to look out for these omens so that they could be interpreted.

The highest point of the Chinese astronomical tradition is often located about A.D. 1280, when Kuo Shou-ching, using the astronomical records of fifteen centuries, improved instruments, higher-order equations, and advanced proto-trigonometry, united his own innovations with the best computational methods that had evolved up to his time to create the basis of his great calendar reform. His approach to calculating the ephemerides remained in use until 1368 and was little changed in the next calendar reform, which provided the basis of practice until the Manchus conquered China in 1644. It may be said that Kuo reached the final plateau of the native tradition. No one went markedly further before the stimulus of the West was felt. Within 200 years after Kuo's death, it is not clear that anyone was still able to comprehend technicalities of his work that he had not spelled out.

What led to this stagnation in astronomy? It has sometimes been explained as only one aspect of a general decay in Chinese science. That there was such a general decay is only a guess. The claim seems to depend largely on mathematical astronomy, the one field of natural science in which progress and regress are straightforwardly measurable, and on analogy with a perceptible slowdown in mechanical invention. But the latter, as we have remarked above, was not demonstrably connected with changes in scientific thought. Nor, for that matter, did it prevent a great deal of innovation in production systems and social organization.

Part of the answer no doubt lies in increasing rigidity of elite attitudes, so that the educated were less inclined to be curious about techniques and less willing to value science as an appropriate pursuit for a gentlemen. At the same time, the official and bureaucratic character of astronomy discouraged its pursuit by the educated merchants and other townsmen who were reinvigorating literature and changing the character of everyday arithmetic over the previous millennium. It has been argued that certain intellectual trends, especially Buddhism and Neo-Confucianism, encouraged the cultivation of the inner self to an extent that ruled out lively curiosity about the external world. One can easily find scientific enthusiasts among both Buddhists and Neo-Confucians, but it is no doubt true that a few talented intellectuals were led away from astronomy by their absorption in religious disciplines.

None of these explanations provides more than part of the answer. For another important part we must look at the influence of foreign astronomical traditions.

INTERACTION WITH FOREIGN TRADITIONS

Attempts to prove foreign origins for Chinese astronomy have not been persuasive, but the translation into Chinese of astrological writings reflecting Indian, Middle Eastern, and Central Asian influences goes back to the third century A.D., by which time the basic goals and methods of the native tradition were well established. Before the middle of the seventh century, after Buddhism had become rooted in China, Indian astronomers were resident in the Chinese capital. The tradition that they applied transmitted in part the methods of Ptolemy, which had reached India after it was conquered by the armies of Alexander the Great. These methods were more reliable than those current in China for the prediction of eclipses. It appears that, from the turn of the eighth century on, the Chinese court depended on resident foreign astronomers for better eclipse predictions than the official system of calendrical computation could provide. A short Indian treatise on the computation of eclipses, mainly lunar, was translated into Chinese by the Indian astronomer-royal at Ch'ang-an in 718 or shortly afterward. It still exists, and has been Englished by the great Japanese historian K. Yabuuti.[7] When the Mongols reunited China under their rule in 1279, their astronomical functionaries were Islamic, from Persia and Central Asia. They disposed a more sophisticated set of computational techniques than their Indian predecessors. They were still performing the same services when the Jesuits began competing with them 350 years later.

The Jesuit missionaries went to China in the early years of the seventeenth century, not for the purpose of propagating astronomical science, but in order to convert the empire to Catholicism. It was the policy of the Society of Jesus to prosyletize from the top down. Matteo Ricci and his colleagues were acute enough to realize that astronomy was the one skill that could give them positions of influence in the Chinese court, positions that had been occupied (with some interruptions) by foreigners for nearly a thousand years. By 1635 Ricci's successors had gained operational control of the Astronomical Bureau after submitting a series of treatises that set out the mathematical foundations of calendrical science. They managed to maintain their position until their order was abolished in the middle of the eighteenth century. Although the period of their work in China was one of rapid transition in European astronomy, they were not obliged to report these changes. Their writings after 1635 were infrequent and appeared mainly when their position was threatened.

Even if the missionaries had intended to disseminate fully the state of the art, the Church's injunction against the teaching of Copernicanism in 1616 and the condemnation of Galileo in 1633 made it impossible for them to do so. Because they understood the importance of Copernicus, they could not resist bringing up his name in their Chinese writings; but regularly when they did so they were constrained to misrepresent what his contribution was. An accurate description of Copernican astronomy was not provided in Chinese until 1760, after *De Revolutionibus* had been removed from the Index of prohibited works. Even then there was no hint of the book's broader implications, nor of how science had changed in the more than two centuries since Copernicus had died.

Successive discussions of the new cosmology by Jesuit writers over nearly a century and a half thus were full of contradictions. As earlier accounts were replaced, the fact that they were irreconcilable with the new ones was seldom noted and never explained. The character of what is often considered the great watershed in European scientific consciousness was thus concealed until the nineteenth century, long after it had become commonplace. It is scarcely remarkable that, as

Chinese and American historians have shown, Chinese scientists late in the eighteenth century were led to believe that the responsibility for the contradictions lay with Copernicus, and considered him a minor and unsuccessful figure.[8]

The Jesuits were more successful in their transmission of mathematical techniques than of cosmology, but the limitation of their objectives again kept them from going very far. Their teachings enabled Chinese to master the mathematical techniques prevalent in Western Europe in the early seventeenth century, including logarithms and fully developed trigonometry. They introduced the rudiments of modern astronomical observation and calculation, and although they never gave Kepler or Newton the attention their theoretical work demanded, they did occasionally introduce the fruit of post-Newtonian observations.

Despite their limitations, the Jesuit astronomical writings precipitated what can only be called a scientific revolution. The best Chinese astronomers of the time (most of them outside the court) responded to the ideas and techniques that the Jesuits described. They adopted new concepts, tools, and methods. They were prompted to change fundamentally their sense of how one goes about understanding the celestial motions. Although their predecessors had been practically unaffected by Indian and Islamic techniques, they set aside traditional numerical procedures in order to use geometric models of orientations in space. They changed their convictions about what constitutes an astronomical problem and what significance astronomical prediction can have for the ultimate understanding of nature. These changes appear to have influenced even main currents in philosophy.[9]

Nevertheless, this metamorphosis of astronomy did not lead to the fundamental changes in thought and society that are naively supposed to be the inevitable outcomes of a scientific revolution. In short, that is because conceptual revolutions, like political revolutions, occur at the margins of societies. The astronomers who responded to the Jesuit writings were members of the educated elite who above all felt a responsibility for strengthening and perpetuating traditional ideals. It is scarcely surprising that they used what they learned from the West to rediscover and master for the first time in centuries the astronomical techniques of their greatest predecessors.

To sum up, an important reason for the declining vitality of Chinese astronomy after 1300 was that the responsibility for the evolution of astronomy remained centered in the imperial court and was largely abandoned to foreign technicians. The Chinese writings of the Jesuits, unlike those of earlier foreigners, became widely available. They were never adequate to help the Chinese advance the forefront of world knowledge; but they stimulated a high pitch of astronomical activity outside the court, although still necessarily within the Chinese elite.

THE ADVENT OF MODERN ASTRONOMY

From the middle of the eighteenth century to the middle of the nineteenth, there was a lull in the transmission of Western science to China. A number of highly educated Chinese, including many of the leading philosophers, studied the old writings of the Jesuits, mastered earlier Chinese writings that threw light on topics that the Jesuits dealt with superficially, and even, in a few cities such as Wuhsi in Kiangsu Province, maintained scientific societies or circles of enthusiasts. As in the Jesuit period, there were no official institutions for the propagation of astronomy among the public.

It was only after the Opium Wars of the early 1840's that Chinese could receive a systematic education in the exact sciences as currently taught in Europe. This time the educators were not individual priests dependent upon the toleration of their hosts. They were Protestant missionaries exempt from Chinese laws, their right to operate missions and schools guaranteed by imposed treaties and enforced by gunboats. They were no longer appealing, as the Jesuits were, to an elite intent on adapting new techniques to traditional ends. The Protestant missionaries educated the poor and people of modest means. Even their richer converts came to them because their children had little chance of conventional success in the old society. The astronomers trained in Western institutions had no reason even to be curious about what their ancestors had done. By 1880 Protestant missionaries, generally working with Chinese, had translated a number of basic textbooks in astronomy, mathematics, and physics and had made them generally available at low prices. Their schools, libraries, bookshops, and other institutions were founded for the purpose of instigating change, not of preserving Chinese civilization.

As the threat of dismemberment by the colonial powers became more imminent, the Chinese government was forced to begin educating modern scientists. In 1866 a department of mathematics and astronomy was added to the T'ung Wen Kuan in Peking, which had previously been a college for interpreters.[10] In 1867 a translation department was added to the Shanghai Arsenal, which had been established 2 years earlier. There Chinese and foreign employees of the imperial government undertook systematically the translation and publication of modern works in science, engineering, medicine, law, and so forth.[11]

As these books were widely distributed, they were eagerly studied by successors of the amateur groups that maintained the tradition that the Jesuit writings had begun. These writings also often played a part in the education of statesmen and reformers, for whom they served as a window on the world. The future lay, however, not with those who saw modern technology as a tool to preserve an empire and a way of life, but with those at the margin of the old society, educated in modern schools and given employment in new institutions.

Beginning about 1900 Chinese astronomers began to emerge who were fully prepared to benefit from advanced training abroad. They were educated in missionary institutions as well as in government universities as these appeared. When the Powers extracted heavy indemnities from China after the Boxer "Rebellion" in 1900, the United States used income from its share to prepare students for technical training abroad. Names of prominent Chinese astronomers who obtained their doctorates in the United States and elsewhere will appear throughout this report.

These twentieth-century pioneers built the observatories that now serve the Chinese people. China's first important observatory was that on Purple Mountain outside Nanking. Its construction occupied the years between 1929 and 1934. During World War II, when Shanghai was occupied by the Japanese, astronomers from Purple Mountain, who moved into the interior, founded an observation station at Kunming in the far southwest. It became Yünnan Observatory in 1972. The Nanking Astronomical Instruments Factory, which supplies all the observatories of China, was founded near Purple Mountain Observatory in 1952, during the Great Leap Forward.

Shanghai Observatory, now the headquarters of the National Time Service, was formed in 1944 out of two small facilities established by French Jesuits in 1872. Peking Observatory, whose four far-flung observation stations are growing in importance, was founded only in 1958. The only observatory that the Delegation did not visit is that of Shensi Province, which was founded in 1967. It is clear that not only the rapid growth of China's observatories, but even the establishment of most of them, are accomplishments of the Communist government since 1949.

ASTRONOMY CONTEMPLATES ITS HISTORY

In China today it is normal for historical studies to be published in astronomical journals, for observatories to have research groups for ancient astronomy, and for modern scientists to be aware of their country's astronomical heritage. This situation contrasts so greatly with that of most other countries that it calls for explanation.

We have already noted that early astronomical records are a valuable resource for modern astronomy. The delegation heard research reports, for instance, that used ancient records of eclipses, comets, sunspots, auroras, supernovas, and earthquakes to yield conclusions of current scientific value. Recent publications cover an even greater spectrum, from ancient latitude measurements to evidence for climatic fluctuations.[12] But this is only one motivation for awareness of history, though undoubtedly the easiest for foreign astronomers to comprehend. Other reasons are related to China's place in the world. For millennia Chinese considered their country the one true center of civilization. Over the last century China has had to make an entirely new place for itself as only one member of a large family of nations. For most of the last hundred years it has been dependent, and looked down upon by foreigners for that reason. The present government is resolved to make it independent, and in doing so has enlisted the full energies of every Chinese citizen.

China's scientific development poses the necessity for adapting a whole sector of society, until a couple of decades ago considered quintessentially Western, to the demands of government policies unlike those found elsewhere. Science and engineering were what one learned from foreigners in order to safeguard oneself against them. This view has gradually been changed by popularizing the history of Chinese science over the last 20 years--not just among scientists but everywhere. Children's books, postage stamps, museum exhibits, school lessons--all have carried the message that science is not European but a world enterprise and that China has been one of the great contributors to that enterprise. Among scientists this consciousness has undoubtedly served to encourage greater continuity between scientific work and political activity.

Enhanced consciousness of Chinese scientific history is not entirely an internal matter. The importance and fascination of the Chinese scientific tradition have long been known in Europe, Japan, and the United States. Scholars in many countries have contributed to understanding it, as well as to making the work of many great Chinese historians of science accessible in other languages. Educated people all over the world are now prepared to respond to new revelations about China's scientific tradition--whether it be innovative applications of the ancient art of acupuncture or the unique archaeological finds that have been appearing without interruption since the 1950's. This heightened interest has meant a small but by no means imperceptible rise in the world's esteem for China. More to the immediate point, it has meant that astronomers all over the world are increasingly aware of the special status of astronomy in China and are less inclined than they might have been to overlook indications that China is rising quickly in the international astronomical community.

REFERENCES

1. See especially *Solid State Physics in the People's Republic of China* (CSPRC Report 1; Washington, D.C., 1976), Ch. 1.

2. Joseph Needham, *Science and Civilisation in China* (7 vols. projected; Cambridge, England, 1954-), III, 458. Our examples and interpretations differ somewhat from those of Needham. His volume is the best available general survey of Chinese astronomy. Readers who wish to explore the technical character of observational and computational astronomy will find particularly useful Ho Peng Yoke, *The Astronomical Chapters of the Chin Shu* (Paris, 1966); Y. Maeyama, "On the Astronomical Data of Ancient China (ca. -100 - +200): A Numerical Analysis," *Archives internationales d'histoire des sciences*, 1975, 25:247-276, 26:27-58; and N. Sivin, *Cosmos and Computation in Early Chinese Mathematical Astronomy* (Leiden, 1969). Studies recommended for orientation in other East Asian traditions are Sang-woon Jeon, *Science and Technology in Korea. Traditional Instruments and Techniques* (MIT East Asian Science Series, 4; Cambridge, Mass., 1974) and Shigeru Nakayama, *A History of Japanese Astronomy. Chinese Background and Western Impact* (Harvard-Yenching Institute Monograph Series, 18; Cambridge, Mass., 1969).

3. "The Origin of the 'Pre-Han Armillary' as Seen from the Silk Manuscript on Planetary Omens" (in Chinese), *K'ao-ku* (Archeology), 1976, No. 143, 89-94, 84.

4. Needham, Wang Ling and Derek Price, *Heavenly Clockwork. The Great Astronomical Clocks of Medieval China* (Antiquarian Horological Society, Monograph 1; Cambridge, England, 1960).

5. In addition to copious Chinese publications on these and other phenomena (and related nonastronomical phenomena such as auroras and earthquakes), reliable summaries in European languages include Ho, "Ancient and Medieval Observations of Comets and Novae in Chinese Sources," *Vistas in Astronomy,* 1962, *5*:127-225; Ho and Ang Tien-se, "Chinese Astronomical Records on Comets and 'Guest Stars' in the Official Histories of Ming and Ch'ing and Other Supplementary Sources," *Oriens Extremus,* 1970, *17*:63-99; and T. Kiang, "The Past Orbit of Halley's Comet," Royal Astronomical Society, *Memoirs*, 1972, *76*:27-66.

6. On sunspots, see Sunspot Compilation Group, Yünnan Observatory, "A Compilation of Chinese Sunspot Records through History and an Investigation of their Periods of Activity" (in Chinese), *Acta Astronomia Sinica,* 1976, *17*(2):217-227; on comets, see the article of Kiang cited in the last note and forthcoming work by Chang Yü-che, Director, Purple Mountain Observatory, Nanking; on supernovas and the origins of radio sources, see David H. Clark and F. Richard Stephenson, *Historical Supernovae* (Elmsford, N.Y., 1977).

7. "The Chiuchihli--An Indian Astronomical Book in the T'ang Dynasty" (in English), in K. Yabuuti (ed.) *Chūgoku chūsei kagaku gijutsushi no kenkyū* (Studies in the history of medieval Chinese science and technology; Tokyo, 1963), pp. 493-538. For a well-informed general discussion of early astronomical transmissions, see Yabuuti, "Indian and Arabian Astronomy in China," in *Silver Jubilee Volume of the Zimbun Kagaku Kenkyusyo* (Kyoto, 1954), pp. 585-603.

8. Hsi Tse-tsung *et al.,* "Heliocentric Theory in China--in Commemoration of the Quincentenary of the Birth of Nicolaus Copernicus," *Scientia Sinica,* 1973, *16*:364-376; N. Sivin, "Copernicus in China," *Studia Copernicana,* 1973, *6*:63-122.

9. Sivin, "Wang Hsi-shan," in *Dictionary of Scientific Biography,* XIV (New York, 1976), pp. 159-168; John Henderson, "The Ordering of the Heavens and Earth in Early Ch'ing Thought," Ph.D. dissertation in history, University of California at Berkeley, 1977.

10. Knight Biggerstaff, *The Earliest Modern Government Schools in China* (Ithaca, 1961), discusses this and other early initiatives.

11. Adrian Arthur Bennett, *John Fryer. The Introduction of Western Science and Technology into Nineteenth-Century China* (Cambridge, Mass., 1967), Ch. 2.

12. Li Kuo-ch'ing, Yi P'ei-jung, and Li Po-t'ien, "Determination of Latitudes in the Yuan Dynasty [1279]" (in Chinese), *Acta Astronomia Sinica,* 1977, *18*(1):129-137, with English summary; Chu K'o-chen, "A Preliminary Study on the Climactic Fluctuations during the Last 5,000 Years in China," *Scientia Sinica,* 1973, *16*(2):226-256.

3

ORGANIZATION AND ADMINISTRATION

SCIENCE AND TECHNOLOGY

Scientific research in China is organized and supported in the context
of a national undertaking to bring about the modernization of agriculture,
industry, national defense, and science and technology by the year 2000.
The present leaders of China view the rapid development of science and
technology as essential for the future welfare and security of the country.
Not only are they well aware of the power of science and technology to
create new industries and to increase the productivity of labor, but
they also understand the potential benefit of theoretical research that
has no obvious immediate practical application. To ensure that science
and technology are properly integrated into the national effort, the
Central Committee of the Communist Party of China (CPC) has established
a State Scientific and Technological Commission (SSTC), which is responsible, with the State Planning Commission, for drawing up a national
program for science and technology as part of a total economic plan for
the country. Thus, the SSTC appears to have the final word on the selection of programs and expenditures for science and technology.

Most institutes for basic research, including the four observatories
we visited and the Chinese University of Science and Technology, are
operated by the Chinese Academy of Sciences (CAS), which provides guide-
lines for organization and research and allocates funding. Many research
institutes, especially those located outside Peking, are managed jointly
by the CAS and by provincial or municipal authorities. Presumably, this
helps the institutes to be more responsive to local problems.

Announcement of the establishment of the SSTC was made in a September 18,
1977, circular of the CPC Central Committee calling for a national science
conference in the spring of 1978. The conference was expected to mobi-
lize the country to work for the four modernizations, and there was full
awareness that the modernization of science and technology was crucial
to the other three. The circular also explained how research institutes,
including astronomical observatories, are organized and administered:

> All scientific research institutions must practice the system
> of Directors' responsibility under the leadership of the party
> committees. It is imperative to install as party committee
> secretaries those cadres who understand the party's policies

and have enthusiasm for science, to select experts or near-experts to lead professional work, and to find diligent and hard-working cadres to take charge of the supporting work.

The role of party committees at all levels is to see that party policies and directives are carried out. Until recently, scientists and technicians were required to devote a large fraction of their time to political studies, but party policy now requires scientific workers to spend at least five-sixths of their working hours in professional activities. The principal party directives to research institutes are now: 1) to excel in science and 2) to eradicate the influence of the Gang of Four.

ASTRONOMICAL OBSERVATORIES

In general, the observatories we visited were administered in accordance with a management philosophy that is profoundly different from that followed in the United States or the West generally. For instance, there is little distinction between scientists and other workers in terms of perquisites and authority. Scientific research is viewed as requiring the combined efforts of administrators (cadres), scientists and technicians, and workers, although the key role of the professionals is acknowledged more now than in the past. An administrator in an observatory in China, as in the West, provides support to the technical staff, but his primary role is to carry out the directives of the party. This does not mean that at any given time party interests may not coincide with those of science, as seems to be the case now. Some administrators we met were trained as scientists, but others seemed to have little formal educational background of any kind.

Peking Observatory has an administrator, Yü Chiang; a director Ch'eng Mao-lan, who is a well-known stellar spectroscopist; and a secretary, Hung Ssu-yi, who was trained as a scientist. Each of the three stations we visited has a separate director reporting to Ch'eng. The director at Mi-yun, Wang Shou-kuan, is an outstanding radio astronomer who was chairman of the Chinese Astronomy Delegation to the United States in 1976. The directors at Hsing-lung, Cheng Yüan-chang, and at Sha-ho, Liang Hsueh-tseng, are both administrators. The observatory is administered by a seven- to nine-person committee with Yü as Chairman. The members are appointed by the Academy of Sciences from a list of nominations proposed by observatory staff, and in conformity with management philosophy of the Communist Party, they are drawn more or less equally from the scientists, administrators, and workers of the observatory. All committee members must presumably also be party members. Special groups, such as women and minorities, may also be represented on the committee. This committee controls the policies and budget of the observatory, while the director is responsible for its day-by-day operation. His role is much like that of an executive officer or a deputy director in an American laboratory. Each of the observatory stations has a Party committee, as do subdivisions that are large enough to include a sufficient number of party members.

During the Cultural Revolution, revolutionary committees were established as administrative bodies in all major institutions in China. They represented the old, the middle-aged, and the young, and were drawn from the mass organizations, the army, and the party. The chief administrators we met in China were also chairmen of their respective revolutionary committees. Since our visit revolutionary committees have apparently dissolved. Institute directors are now individually responsible for the work of their institutions and report to the party committee.

The directors of Yünnan and Shanghai Observatories are, respectively, Wu Min-jan, a scientist, and Chou Tsun-po, an administrator. The director of Purple Mountain Observatory, Chang Yü-che, is a well-respected astronomer who received his Ph.D. in the 1930's at Yerkes Observatory as a student of G. van Biesbroeck. He has been President of the Chinese Astronomical Society for at least 20 years.

We were told that the director's role, in addition to managing the observatory, is to advise the committee on scientific matters. If a disagreement develops between the director and the chief administrator, the director has the right to appeal up the line to the CAS, which may, if it wishes, take up his case at appropriate levels of the party.

An advantage of the Chinese system of administration is that observatory administrators, through their party connections, have access to decision-making levels of the government. A disadvantage is that administrators and party committees who are uninformed about science may have difficulty in judging the relevance of new fields of science to national needs.

4

INSTITUTIONS FOR RESEARCH
AND EDUCATION

The principal organizations that conduct astronomical research and education in China are either institutes within the Chinese Academy of Sciences (CAS) or departments within several of the universities. A brief description of these organizations follows. The principal instruments, research programs, and staff are summarized in Appendix A.

CAS INSTITUTES

The principal astronomical institutes of the CAS, in the order that we visited them, are Peking Observatory, Yünnan Observatory in Kunming, Shanghai Observatory, Purple Mountain Observatory in Nanking, and the Nanking Astronomical Instruments Factory.

Peking Observatory

This new institute of the CAS was begun in 1958 during the years of the Great Leap Forward. The five research departments of (a) solar physics, (b) stellar physics, (c) radio astronomy, (d) astrometry, and (e) latitude service are centered in four field stations, the first three of which we visited:

Sha-ho Station, 20 km from Peking, is the center for solar physics. Construction of the station began in 1958. It has a staff of 100 researchers and assistants plus 80 support technicians.

Hsing-lung Station, 100 km northeast of Peking, is the stellar physics center. Construction of the station began in 1964 and was completed in its present form in 1968. It has a staff of 38, of which 25 are astronomers and 13 are technicians. This is to be the site of the 2-m telescope

Mi-yun Station, 100 km northeast of Peking, is the radio astronomy center. Construction of the station began in 1967. It has a staff of 80.

Tientsin Station, the principal latitude station for China, was the first station of Peking Observatory to be built. Construction began in 1957.

Yünnan Observatory, Kunming

The site of Yünnan Observatory was originally an observing station of
Purple Mountain Observatory and served as its transported headquarters
during World War II. The optical parts of the observatory's 60-cm reflector were stored here during Japanese occupation of Nanking. The
observatory was set up as an independent institute of the CAS in 1972,
and construction on the present very expanded plan was begun in 1975.
Upon completion, Yünnan Observatory will have the largest astronomical
physical plant in China.

The observatory has five research departments: (a) solar physics,
(b) astrophysics including stellar physics, (c) astrometry, (d) celestial
mechanics, and (e) technical group (computers, data processing, electronics, mechanical engineering, optical shop, etc.). We were told
that there are now approximately 100 people in each group, a total of
500 researchers and support personnel.

Shanghai

The original observatory, founded by French Jesuits in 1872, was centered
at two observing sites, one in Shanghai and the other 30 km distant
(Zo-se Station, which we did not visit). Until 1950, when the stations
became part of the CAS, the primary work at the Shanghai site was rudimentary time determinations. All equipment now in use there has been
installed since 1949, and most postdates 1950. However, we were informed
that the original 40-cm astrograph of 7-m focal length, which dates from
1900, is still in use at Zo-se and is currently used for second and
third epoch proper-motion work. The observatory's three research sections are (a) time, (b) atomic and shock-wave time standards, and
(c) photographic astrometry and astrophysics. The present staff of
research workers plus support staff is approximately 100.

Purple Mountain Observatory

This is the oldest observatory in modern China. Construction began in
1929 and was completed as to physical plant in 1934. In 1937 the
Japanese occupation began and astronomers moved to Kunming. The pre-
1949 staff was small (10 astronomers). After 1949 the observatory became part of the CAS with greatly increased staff that now numbers 150
workers plus support personnel. The observatory's eight research sections are (a) solar physics, (b) stellar physics, (c) radio astronomy,
(d) planets, (e) ephemeris (almanac), (f) time service, (g) ancient
astronomy, and (h) satellite tracking.

Nanking Astronomical Instruments Factory

This factory was established during the Great Leap Forward in 1959 in
southeastern Nanking as a unit of the CAS to construct telescopes and
a variety of other optical research instruments for Chinese observatories

and observing stations. The factory now has more than 300 workers and technicians, including mechanical, electronic, and optical engineers and designers. It occupies a large tract of land where 10 separate modern buildings house the optical, mechanical, and electronic workshops. But even this very large present facility is apparently not adequate for the needs, as new expansion of the factory was evident in the form of ground-clearing for additional buildings.

The factory is engaged in the design and construction of complete astronomical instruments for observations of artificial satellites and the Sun, for astronomical time determination, and for stellar astronomy. The design of instruments involves active cooperation between the engineers of the factory and the astronomers that will use the instruments. More than 20 different types of small- to moderate-size instruments have been produced so far. For observation of artificial satellites, they have built Schmidt telescopes and optical printing theodolites and are now manufacturing a large satellite-tracking camera, which we saw mounted in the factory. For solar work they have built horizontal solar telescopes for Yünnan and Purple Mountain observatories and are now making a new chromospheric telescope of increased resolution. For time service and latitude determination they have built photoelectric transits (astrolabes) of types 1 and 2, and a photographic zenith tube, as well as a type 3 photoelectric transit instrument. For astrophysics they have produced a 60-cm reflecting telescope and are now working on a 2-m telescope. For the chromospheric telescopes, they are building a new Hα filter consisting of 4 calcite and 26 quartz components with a half-band width of 0.5 Å. A more complete description of the factory and its activities is given in Chapter 6, where its function relative to construction of the 2-m reflector is discussed.

UNIVERSITIES

In addition to these institutes of the CAS, astronomy is conducted at three universities. We had direct information on two universities during visits there, and information on the third during the symposia.

These institutions are: Peking University, Nanking University, and the Chinese University of Science and Technology.

Peking University

Established in 1898, this university has 10 departments of science (including mathematics, mechanics, physics, chemistry, and geophysics); 7 departments of liberal arts (including history and philosophy); and 3 departments of Western languages (English, French, and Spanish).

The President of Peking University is Chou P'ei-yüan. The university has a faculty of 2,700 and a student enrollment of 7,000, of which 130 are from 37 countries.

The astronomy section, which is part of the geophysics department, has a faculty of 17 and a student enrollment of 52. In February 1978, 20 new astronomy students were expected, in addition to the 52 returning

students. The principal teaching equipment is a 2-m solar radio telescope, but apparently there is no optical equipment.

Nanking University

Nanking University began in 1902 as a normal college. Prior to 1949
its name changed many times--sometimes to Southeastern University or
Central University--but since 1949 it has been known as Nanking University.
Before 1952 it had seven colleges: (a) arts, (b) law, (c) agriculture,
(d) medicine, (e) normal education, (f) science and technology, and
(g) engineering technology. There was no astronomy department.

In 1952 all universities in China were reorganized, after which
Nanking University retained only the schools of arts and of science.
At that time the astronomy departments of Canton and Shantung Universities
were moved to Nanking to be close to Purple Mountain Observatory. Today,
in addition to astronomy, there are eleven departments in the schools
of arts and of sciences, including Chinese, history, politics, geology,
geography, languages, mathematics, meteorology, physics, chemistry,
and biology.

The university's student enrollment was 700 in 1952 and had reached
6,000 in 1965-66, but was reduced during the Cultural Revolution. In
1977 there were 3,000 students, with housing and food facilities for
6,000. The university has 1,400 faculty members.

Professor Tai Wen-sai is head of the astronomy department, which has
50 faculty members and 120 undergraduate students in astronomy with
30-40 graduating students in astronomy per year. The four principal
divisions within the department are (a) stellar and solar astrophysics,
(b) radio astronomy, (c) astrometry, and (d) celestial mechanics. The
chief activity of the department is training people for the national
observatories of the CAS. Subjects taught in astronomy are:

Astrophysics (stellar atmospheres and interiors)
Radio astronomy
Physics of solar activity
High-energy astrophysics
Structure and evolution of galaxies
Earth's rotation, polar motion, etc.
Fundamentals of celestial mechanics
Orbits of satellites
Astrometry

Research in many of these fields is carried on by a number of the
faculty. The department will soon begin accepting research students
(comparable to our graduate students), in line with the new emphasis on
the training of experts laid down by the Chinese leadership.

Chinese University of Science and Technology

The Chinese University of Science and Technology was founded in 1958
in Peking and moved to Hofei in Anhwei Province during the Cultural

Revolution. We did not visit the university, but met a number of faculty members at the Peking and Nanking symposia. The physics department does work in quasars, high-energy astrophysics, density-wave theory of galaxies, and cosmology, etc. We did not learn the number of students or faculty, but during our visit the Central Committee in Peking announced the start of an intensive program to train graduate students (1,000 to begin residence) at the university beginning in 1978.

5

RESEARCH INSTRUMENTS
AND PROGRAMS

SOLAR PHYSICS

Observational solar research in China is devoted principally to studies of solar active regions and solar flares at four locations: Sha-ho Station of Peking Observatory, Purple Mountain Observatory, Yünnan Observatory, and Shensi Observatory. In the optical region of the spectrum, solar images in white light and in Hα are routinely photographed with small telescopes, and spectra with resolution of about 100,000 are recorded at Yünnan and Purple Mountain observatories with 30-cm horizontal telescopes and multichannel spectrographs, which are also operated as spectroheliographs and will soon be equipped with Leighton-type magnetographs for the measurement of magnetic fields stronger than 40-50 G. Research instruments for the radio observation of the Sun are highlighted by a fine interferometric array of 16 9-m dishes operating at 450 MHz at Mi-yun and include also a number of small instruments for patrol work at 3 and 10 cm at Sha-ho, Purple Mountain, Shensi, and Yünnan observatories.

Under the best normal seeing conditions (at Yünnan Observatory), the optical spatial resolution is 1"-2" in very short white-light exposures and 3"-5" in Hα images. The equipment seems clearly designed for research on solar active regions and the research reports we heard confirmed this impression. The spatial resolution is insufficient for studies on the quiet Sun, and, moreover, the spectral resolution is not high enough for work on the profiles of weak lines. Thus, at Peking Observatory, Shih Chung-hsien and his collaborators are studying the relations between large flares and the magnetic configurations of sunspot groups as well as the longitudinal distribution of solar activity regions that produce proton flares. At Yünnan Observatory, Ting Yu-chi, Chou Yün-fen, and others are investigating relations between the structure of sunspot groups and the frequency and magnitude of flares within them. One of the principal goals of these studies is the prediction of flares, especially those associated with proton events. We were, in fact, informed that solar patrol observations of radio emission, flares, and sunspots from Sha-ho, Yünnan, and Nanking were all sent routinely to Peking for predictions of solar activity, and in addition that Purple Mountain Observatory made predictions from its own data.

Instruments

A complete listing of solar instruments and principal research programs is given in Appendix A. We add here some further relevant comments about each observatory.

Sha-ho Station   The principal instrument for research at this station is a 60-cm coude reflector, which forms a 35-cm solar image on the slit of a plane grating spectrograph (Figure 4). The grating has a ruled area 12 × 15 cm, ruled 600 lines/mm. It was manufactured at Changchun in Kirin Province. On the morning of our visit, the seeing appeared to be fairly good, about 3". The spectrograph is used primarily to make visual measures of the magnetic fields of sunspots, which are used in studies of the origin of flares and flare forecasting.

The Hα chromospheric telescope is used with a Lyot-type filter, transmitting a wavelength band with width 0.5 Å at half intensity. The 2-cm

FIGURE 4   Sha-ho 60-cm solar telescope.

diameter image is photographed on film made in China. Pictures may be taken up to a speed of about four per minute during flares. The white light telescope on the same mounting gives an 8-cm image, which is photographed on glass plates imported from Germany.

Yünnan Observatory Three principal instruments for solar research are in operation at Yünnan Observatory. The 12.5-cm diameter Zeiss refractor, which forms an image of diameter 17.7 cm, is used for the mapping of sunspots and the photography of fine structure in sunspots. The solar image is projected on a white card and photographed with a hand-held 35-mm camera at 1/1,000 s when the observer perceives moments of good seeing. We were told that an image resolution of 1"-2" is obtained routinely by this method.

Patrol observations in Hα are taken with a chromospheric telescope, which was manufactured at the Nanking Astronomical Instruments Factory in 1966. The aperture of the telescope is 14.5 cm, and the image diameter is 1.55 cm. The telescope is used with a Lyot filter of French manufacture and a movie camera made in Nanking. Exposure times can be varied between 0.5 and 5 s at intervals of from 4 s to 12 min. Data on bright flares are provided with normal resolution of 3"-5", which reduces to about 2" under the best seeing conditions. Observations are carried out on 310 days per year for an average of 9 h/day and up to a maximum of 13 h/day.

In 1966-68 the Nanking Astronomical Instruments Factory and Purple Mountain Observatory collaborated on the design and construction of telescopes and multiple-band plane grating spectrographs for Yünnan and Purple Mountain observatories. The instruments were designed for the study of flare spectra during periods of high solar activity and were built in time for observations during the 1968-69 solar maximum. In both telescopes, coelostats and second flats of 40-cm aperture are used to feed imaging mirrors 30 cm in diameter.

The focal lengths of the Yünnan and Purple Mountain observatory mirrors are 16 and 12 m, respectively, giving solar images 15 and 11 cm in diameter. The Yünnan telescope may also be converted to an off-axis Cassegrain system with an effective focal length of 45 m, which forms an image 40 cm in diameter for the observation of fine details of the solar surface. The gratings were produced at Changchun and have ruled areas of about 10 × 10 cm with 600 grooves/mm. The Yünnan Observatory spectrograph has 10 cameras, each of which is used to photograph a narrow band of wavelengths centered on a strong Fraunhofer line. The focal lengths of the cameras are about 7 m, giving a linear dispersion in the second order of about 1 mm/Å. Similarly, the Purple Mountain Observatory spectrograph photographs 9 wavelength bands simultaneously, but the telescope does not have a Cassegrain arrangement. The cameras may be operated automatically by electromatic control with variable exposure times. Between 8 and 20 spectra are photographed on a single plate, 18 cm long. At Yünnan, the spectrograph is used as a spectroheliograph for rough mapping of magnetic fields in sunspots. The instrument is also used for the study of line profiles and flare spectra. The resolving power of the Purple Mountain Observatory spectrograph in the second order is 90,000, as compared with 120,000 for that of Yünnan Observatory.

Purple Mountain Observatory   The chromospheric telescope of Purple
Mountain Observatory was made by the Secasi firm in Bordeaux in 1957-58.
A number of these telescopes were built for observatories participating
in the International Geophysical Year program. The telescope has a diameter of 16 cm and a focal length of 1.4 m, giving an image 1.3 cm in
diameter. The Hα filter has a band pass of 0.75 Å. In normal operation
the camera takes four pictures per minute, which is reduced in the patrol
mode to one every 15 minutes. Under normal seeing conditions, the image
resolution is 3"-5" and may reach 2" under the best conditions. Ilford
film is used, and calibration is provided by eight density steps, which
is adequate for the normal active sun but does not provide a sufficient
dynamic range for flares.

Shensi Observatory   Shensi Observatory, which was founded in 1967, is
one of five independent observatories. It has a staff of 150 scientists
and technicians. According to the Nanking astronomers, the observatory
has a small dish for solar radio astronomy at 3.2-cm wavelength, a
transit instrument used mainly for time service, a photoelectric astrolabe, and instruments for satellite tracking.

Research Programs

In the People's Republic of China, even more than in the West, theorists
prefer to work on exotic subjects such as relativity, cosmology, and
spiral structure of galaxies and to neglect fields like solar physics.
We heard few reports of theoretical solar investigations and none deal-
ing with such forefront problems as global oscillations, the origin and
evolution of large- and small-scale magnetic fields, differential rotation, the origin of flares, coronal and interplanetary dynamics, etc.
During visits to the three principal observatories engaged in solar
research, we heard two theoretical reports on solar topics. Cao Tian-yun
of Purple Mountain Observatory described a calculation of the geometry
of magnetic fields in a coronal condensation observed during the total
solar eclipse of September 22, 1968. Liu Hsü-chao of the Mi-yun Radio
Observatory reported on a "Mechanism of Solar Type I Bursts," a paper
in which the bursts are explained as a consequence of parametric coupling
between whistler and Alfvén waves. Both studies were competently performed.
    A modest program of theoretical and analytical work on the Sun is
also in progress at Peking and Nanking Universities.

Nanking University   Yao Pan and eight collaborators in the astronomy
department, Nanking University, have derived a theoretical temperature
model of the lower and middle chromosphere in which heating is assumed
to occur by shock waves formed from acoustic waves. Based upon the
Harvard Smithsonian Reference Atmosphere, the most probable periods of
shock waves heating the low and middle chromosphere are found to be 20-
30 s and 50-100 s, respectively. For stable active regions the probable
period of shock heating of the lower chromosphere is about 300 s. There
is some evidence for the presence of oscillations in this frequency range

in the observations made with equipment on OSO-8. The work shows that the authors are well acquainted with Western literature in the field.

Peking University   During our visit to the astronomy section of the geophysics department of Peking University, we heard reports by P'eng Ch'iu-ho on "The Stable Confinement of Fast Electrons in the Region of Type I Bursts at Meter Wavelengths"; by An Ching-chu and two others on "A Possible Design for a Millimeter Wave Solar Radio Heliograph Using Aperture Synthesis"; by Yang Hai-shou on "The Wilson Effect in Sunspots," in which he used observations made at Yünnan Observatory; and by Chou Tao-ch'i on "The Origin of the Solar Basic Magnetic Field and Sunspot Activity."

Yünnan Observatory   Important contributions to solar physics are being made through the study of ancient records of sunspots and aurorae. At Yünnan Observatory, Chang Chu-wen and the "Group on Arrangement of the Chinese Ancient Sunspot Records" have carried out an autocorrelation analysis of 112 sunspot records of observations made during the period 43 B.C.-A.D. 1638. Presumably, the spots were observed with the naked eye through haze in the early morning or late afternoon. The 11-year cycle was found to be quite distinct, and two longer periods of about 65 years and about 250 years were also reported. A separate analysis was also made for the period of the Maunder minimum, 1640-1715, studied by J. Eddy. Using Eddy's numbers for the period A.D. 1642-1715, Chang reports signs of an 11-year cycle with a period of 11.0 ± 1.9 years. Chinese records during the Maunder minimum suggest Eddy's numbers are low. For example, in 1656 there was at least one large spot of area $2,000 \times 10^{-6}$ of the Sun's visible disk. The results indicate that the 11-year cycle is not a recent phenomenon and in fact has existed over the last 2,000 years.
    Similar analyses of auroral (207 B.C.-A.D. 1517) and earthquake (70 B.C.-A.D. 1643) records were reported by Lo Pao-yung and Li Wei-pao and lead to the conclusions that geophysical phenomena such as aurorae and earthquakes are closely related to solar activity and that the 11-year cycle deduced from auroral and earthquake data has also existed in the last 2,000 years. No reasons were given for expecting a physical connection between solar activity and earthquakes.

Peking Observatory   Finally, Peking Observatory astronomers have taken measurements of the solar spectral energy distribution in the region from 0.60-1.20 μ with a monochromator at a site at an altitude of 5,000 m in the Himalaya Mountains in South China. We were told that the results obtained agree with earlier work by H. L. Johnson. The availability of such high-altitude sites could be an important asset to the Chinese if they should wish to establish a coronagraph station, which would be of considerable practical value, for example, in the forecasting of geomagnetic disturbances.
    Although solar research in China until now has had a strong utilitarian flavor, the recent renewed emphasis on the importance of basic research suggests that a more diversified program may be in prospect. The questions asked during discussion periods indicated that the Chinese were

at least thinking about new directions for future solar research. For example, the solar astronomers in Peking, Kunming, and Nanking were interested in such matters as comparative image quality in evacuated and unevacuated solar tower telescopes, the factors determining solar seeing, and the detailed design of the Kitt Peak solar vacuum telescope and magnetograph. They also requested and received lectures on recent developments in solar physics in the United States covering such topics as small- and large-scale magnetic fields and global oscillations derived from measurements of both velocity fields and solar diameters. All of them had carefully read the Western literature and had many questions that revealed their eagerness to "learn from others" in planning the construction of new facilities. For example, the Kunming astronomers were particularly concerned about image disturbance in the Kitt Peak solar spectrograph "when the train went by." We explained that no trains pass within 80 km of Kitt Peak, which in any case provides bedrock foundations, whereas Yünnan Observatory, where trains do pass nearby, is built on laterite.

SOLAR SYSTEM

Research Programs

Chinese solar system research is concentrated at Nanking, primarily at Purple Mountain Observatory, but also at Nanking University. Several dozen astronomers, assistants, and students are involved, and about 10 percent of the recent articles in the *Acta Astronomica Sinica* are from this group. At least five areas of solar system research, apart from the Earth and Sun, appear to be active, as noted below.

Purple Mountain Observatory   A program at Purple Mountain Observatory using the Schmidt and the Astrographic Telescopes is devoted to improving orbits of some faint comets and asteroids. Plates taken for this and other purposes such as photometry, are also scanned for new objects; several interesting short-period comets and Mars-approaching asteroids have been discovered.

Several telescopes equipped with photoelectric photometers have been used for such projects as studying changes in the Eros light curve, observing connected phenomena (occultations and eclipses) of Jovian satellites, and the Uranus stellar occultation that led to the discovery at a number of observatories of the multiple rings. In each of these cases a sophisticated analysis of the data has been made, including extensive use of computer modeling and data filtering in order to try to extract the maximum information from low-signal records (the actual significance level of some of the presumed Uranus-associated events is not clear).

The Nanking group's modern approach to dynamical astronomy applies general analytic theories, improved special theories, and computers to calculate specific cases of interest such as comet and asteroid orbits, shapes of comet tails, and orientation of the pole of Eros. They have also done some work on the theory of artificial satellites, but the center of activity in that field may perhaps shift to Kunming as the large new computing and satellite tracking facilities are completed there.

Nanking University   The theory of the origin of the solar system has been carefully reviewed at Nanking University, involving a study of essentially all cosmogonic theories in the literature. The conclusion, consistent with most Western views, is that a highly probable mechanism was the collapse of a nebula with ice and dust condensation, planetesimal accretion, and planetary consolidation, with rotation imposed by the impacting planetesimals.

Meteorite falls and finds represent a fine research opportunity for so densely settled a country as China. Such work is in progress, including not only pursuit of falls and mineralogical and chemical analysis of finds, but also potassium-argon dating.

In summary, solar system astronomy is a specialized but healthy research field in China today, with what appears to have been strong early encouragement of observational and dynamical research, particularly by Chang Yü-che, director of Purple Mountain Observatory, and of theoretical research by Tai Wen-sai, head of the astronomy department at Nanking University. Observational programs have been chosen that are appropriate to the telescope apertures available, and powerful analytic tools are being applied to the data by Ch'en Tao-han of the Purple Mountain Planetary Laboratory, among others.

Possible Future Research

One area where important new results might be achieved with the present relatively modest equipment is the search for small asteroids crossing the Earth's orbit. Recent American studies show that hundreds to thousands of these must exist that are large enough to detect on the fleeting occasions when they are relatively near the Earth. A handful should have orbits differing so little in energy from the Earth as to be feasible for close study and perhaps sample return within one or two decades, and possibly someday even for capture and utilization. The search would involve use of Schmidt cameras and of the large new satellite-tracking cameras being built at the Nanking factory. Use of the fastest available film coupled with immediate development and inspection would permit rapid re-observation both locally and internationally and thus ensure orbit determinations before the fast-moving objects could get away. The yield each year would be low, but some of these objects could be very important in the future despite only very rare favorable opportunities for detection.

TIME SERVICE AND RELATED ASTRONOMY

Chinese work on time measurement and standards is directed towards providing China with an independent and self-contained system for determining time and polar motions. However, Chinese astronomical time determinations are regularly compared with other internationally available standards and measurements. Precision of measurement is said to be comparable with the best available elsewhere. In any case, the entire complex for time measurement and related astrometry seems to be one of the more extensive astronomical efforts, and overall results are probably closer to the forefront of performance than those in other broad areas of observational astronomy.

The amount of work in the field of time standards and astrometry is impressive, being a major part of almost every observatory. The astronomy department of Nanking University graduates about 10 astronomers per year who have done undergraduate theses on a transit instrument similar to those installed at observatories. Presumably this particular specialty will play a relatively less important role as astronomy is expanded, as new higher precision techniques such as long baseline radio interferometry and satellite ranging are introduced, and as there may be less emphasis on applied work.

Instruments

Shanghai Observatory has primary responsibility for Chinese time standards and receives and analyzes pertinent information from all other locations in China. A full list of observatories and stations involved in time measurement and the observing equipment used at each appears in Appendix A. The instruments and the accuracy obtained are briefly described below.

The Danjon visual astrolabe is made in France and hence is well known from other sources. As many as 60 stars may be observed per night, and at Shanghai 120 nights per year are adequately clear for observing. The error in measuring the Earth's mean rotation rate is as small as 3.5×10^{-3} s/year. The precision of measurement is said to be somewhat poorer than that of the photoelectric astrolabe described below. Hence, its use is expected to be discontinued in favor of the photoelectric astrolab after somewhat more than the present 2 years of comparison.

Photoelectric astrolabes have been made since 1971 by the Nanking Astronomical Instruments Factory, type I having been used at an observatory since 1973 and type II at Shanghai Observatory since 1974. The factory is now working on a type III instrument. Distinctions between these types are not clear and will not be made here. This instrument was designed in collaboration with Shanghai and Peking observatories. It involves a 20-cm aperture Ritchey-Chretien telescope of 2.4-m focal length and evacuated to minimize refractive problems, angle mirrors of fused quartz, a mercury horizon, and photomultiplier detection of the starlight. The latter is modulated by a chopping wheel and synchronously detected. The two star images move along a grating, with rulings separated by about 2 s of arc. The difference in time when each image is cut by the edge of a ruling is automatically recorded. Twenty such comparisons are made immediately before image convergence and 20 immediately after, and the results are averaged. The precision of such an average is said to be somewhat better than that for either the Danjon astrolabe or the transit telescope, which would make it slightly better than $\pm 10^{-2}$ s of time. Latitude accuracy is ± 4-5×10^{-2} s of arc. The limiting stellar magnitude that can be used is 7, about 1 magnitude weaker than what is obtainable with the Danjon instrument. As many as 120 stars a night can be measured.

The Zeiss transit instrument, the "broken type" with a 10-cm aperture and 1-m focal length, was modified in China for photoelectric detection. It has a precision for a single measurement of $\pm 13 \times 10^{-3}$ s of time. At Shanghai Observatory a series of measurements are made every 2 weeks, with an accuracy for the average of the series of $\pm 3 \times 10^{-3}$ s of time when compared with the yearly average rate. "Secular" variations of the same magnitude are obtained at Sha-ho.

We did not see a zenith tube, but a photographic zenith tube recently made by the Nanking Astronomical Instruments Factory is described in *Acta Astronomica Sinica* (vol. 18, p. 38, 1977) and was described to us at the factory. It has a 26-cm aperture and f 15.86 optics, and the tube is evacuated to minimize refraction problems. It has a photoelectric tick rack, involving a scale, optical magnification, and a photoelectric detector to provide accurate timing, and also a laser interferometer for checking the rotational position of the plate holder. One now installed at the Tientsin Station gives precision of $\pm 13 \times 10^{-3}$ s of time for one star. Tests at the factory show an accuracy of 0.13 arc sec in latitude. Tientsin has previously had a "small" zenith tube, of unknown type.

Clocks

Whereas quartz crystal clocks appear to be used as standard local clocks at most observatories, a variety of atomic clocks are under development or already available. At Sha-ho and Nanking, the average time readings of several quartz crystal clocks are used. While Sha-ho has a rubidium atomic clock made by Rohde and Schwartz of West Germany, which has an accuracy about 10^{-11}, it is not used as a local standard because there is not a second or third one against which to compare it. Shanghai's large array of atomic clocks includes five rubidium clocks, some made in Shanghai and some in Peking at Yüan-tzu p'in-lü piao-chun. These are said to have a stability of 10^{-12} and a variation among themselves of about the same magnitude, which compares favorably with results for rubidium clocks in the United States. While their local clock is run from a quartz crystal oscillator, it is regularly compared with the rubidium systems, which serve as local standards.

Much earlier the group at Shanghai had built ammonia maser clocks, but these are not now in use. In 1970 the development of hydrogen maser clocks was initiated and two working hydrogen masers have been on hand since 1973. Their stability over modest times was said to be 10^{-13}, but over longer times there are variations of $1-2 \times 10^{-12}$, believed to be associated with temperature fluctuations. The temperature at some point on the quartz hydrogen container about 30 cm in diameter is stabilized to ± 0.03 K, but it seems likely there are larger variations from point to point. Two additional hydrogen masers are being assembled, and it is expected that hydrogen masers will ultimately be used as standard local clocks. The Sha-ho Station expects to obtain a hydrogen maser, presumably from Shanghai.

While there is no work on caesium clocks at these observatories, caesium systems are being worked on at the Institute of Metrology in

Peking. This institute has responsibility for developing standards and calibrating instruments for laboratory and industrial use. We understand from other sources that about 10 caesium clocks have been ordered for China from Oscilloquartz in Switzerland.

Observatories other than Shanghai provide their time information to Shanghai and also compare their time signals with Shanghai, in some cases also checking against the Japanese station JJY or other foreign broadcasts. Sha-ho time can, for example, synchronize with Shanghai to 10^{-6} s by using a TV display. However, their time is not kept more accurately than about 10^{-2} s. Sha-ho and other stations provide locally needed time signals, such as to broadcast stations. The more accurate Shanghai time signals are generally used for scientific work.

The Shanghai time station BPV broadcasts UT_1 on 9,351 kHz and 5,430 kHz and UTC on 5, 10, and 15 MHz. In addition, the station XSG broadcasts UT_1 on 458, 6,414, 8,502, and 12,871 kHz. The international time signals of WWVH, JJY, and RWM and information from BIH are regularly monitored. Shanghai UTC signals are found to vary about 10^{-3} s with respect to BIH, and are adjusted when necessary to stay within this tolerance. Loran signals from a station in Japan are also used to some extent to synchronize with other Chinese observatories. Shanghai's precision for UT_1 is also 10^{-3} s and the rate accuracy of UTC is 10^{-10}.

For future improvement in the precision of time determinations, Shanghai astronomers are pursuing two main paths: use of a long baseline radio interferometer and more experiments on hydrogen masers. Wu Lin-ta, an electrical engineer by training, is especially interested in a radio interferometer and has begun work towards constructing one. He was also very interested in plans of this type in the United States.

Studies on Polar Motion

Apparently Chinese astronomers have not previously made their measurements on the wander of the polar axis available to other countries, but have done substantial statistical studies on these and on comparison with foreign measurements. They now seem to be in a position to publish results, and we were given reports of these studies printed by Shanghai Observatory specially for our visit. In general, they point out inconsistencies clearly beyond random errors in the measurements of polar axis motions made in various places. The most detailed comparisons we were shown were between their measurements and those in the United States. They conclude that a substantial part of this apparent motion of the pole must be attributed to local variations in the direction of gravity at the observatory stations, since every instrument used so far is oriented with respect to gravity.

SATELLITE OBSERVATIONS AND ORBITAL DETERMINATIONS

Rather high priority has evidently been given to measurements and orbit determinations of artificial satellites. A number of students are being trained in this area, computers are put into service for orbit calculations, and instrumentation is fairly advanced. Satellite orbital determinations are a significant part of the work at Purple Mountain and Yünnan observatories, and one can assume that there are several tracking stations in addition to those we visited. However, we were not told of any program of space astronomy in China, and the observatories presently do no work on satellite construction or instrumentation. Astronomers hope to do satellite astronomy in the future, but apparently no schedule for such work has been set.

Research Programs

Purple Mountain Observatory   Purple Mountain Observatory carries out work on satellites, asteroids, comets, and meteorites. There are about 30 people working on artificial satellite orbits and associated theory, and on compiling an almanac. Astrometry and celestial mechanics are two of the observatory's four principal divisions, and there has been theoretical work on satellite orbits and their calculation since 1958. In 1964 the Schmidt telescope, of focal lenth 80 cm and 7.6^0 field of view, was set up, and at the same time the observatory began to compile ephemerides and an almanac. This telescope made orbital measurements on China's first two satellites, in 1970 and 1971, and is also being used for asteroid study and other astronomical problems. A rotating focal plane shutter can provide a time precision of 10^{-4} s, but the actual accuracy for measurement of a satellite track is 10^{-3} s of time and 1 sec of arc. There is also a tracking theodolite of 15-cm aperture at the observatory, which we did not see, and the observatory is now setting up two-frequency radio Doppler work on satellites. For orbit computations, for the past 10 years, Purple Mountain Observatory has had a computer with a speed of 25,000 operations/s. It has recently acquired one of China's TQ16 computers, with a speed of 120,000/s, but this is yet to be installed and become operational.

Nanking University   Astrometry and celestial mechanics are also two of the four divisions of the astronomy department at Nanking University, which graduates 30-40 astronomers per year and is China's principal institution for academic training in astronomy. Theoretical work there includes the study of satellite orbits.

Yünnan Observatory   Yünnan Observatory in Kunming seems destined to be the primary center for work on satellite orbits and cataloguing. Celestial mechanics and astrometry comprise two of the five divisions of this observatory. A new building with about 30,000 ft^2 of floor space, including space for a telescope, is being built at the observatory for celestial mechanics. There is also an expanding microwave effort, but the extent to which this may be used in connection with satellites is not known. Lin Chao-chü, principal scientific spokesman for the observatory, says that there is no work on planets at Yünnan and that the observatory undertakes work on satellites as an assigned applied job which has no specific connection with its astronomical work. Part of this assignment, he says, is to do for China what the Smithsonian Observatory does for the United States in maintaining information on satellites in orbit. While we did not see the satellite-tracking equipment at Kunming, we were told that a group there observed China's satellites, which are tracked visually with a Chinese-made instrument. At Kunming there are now a satellite-tracking theodolite and a small tracking camera. We did see there a TQ16 computer to be used for orbit computations. It was recently installed in a building constructed by the astronomers themselves (!) because they were so eager to begin using the computer. The computer is equipped with two magnetic drums and a 32-K core memory.

Shensi Observatory   Shensi Observatory near Sian, which we did not visit, has a satellite-tracking telescope. No satellite work at other institutions came to our attention.

Instruments

The two most impressive indications of the seriousness and vigor with which the Chinese are undertaking satellite observations were the large and rapidly growing facilities at Yünnan Observatory and the emphasis that has been put on tracking instruments at the Nanking Optical Instruments Factory. This factory makes tracking theodolites, of which the Purple Mountain and Yünnan instruments are perhaps examples. It was their satellite-tracking camera, however, a Schmidt of 85-cm diameter primary, which was most striking (Figure 5). The designer, Tsu Chi-tao, explained its parameters to us.

The guiding system of this tracking camera has a 20-cm aperture, an apochromatic lens, and a fixed eyepiece. The Schmidt camera has an effective aperture of 60 cm, a doublet correcting lens, a 70-cm focal length (f/1.17), and a 5° by 10° field of view, the direction of the largest field of view being oriented in the tracking direction. The system is a rather massive and impressive device. It has four axes, so that one axis may be pointed perpendicularly to the orbital plane for easy tracking. There is a slit shutter at the focal plane that obscures a satellite track five times during an exposure for accurate timing. The camera may be loaded with a roll of film as long as 15 m; the film is vacuum supported during exposure. An exposure time of 1/4, 1/2, 1, or 2 s may be selected. The telescope can be operated in either a stationary or a tracking mode, or in a mode designated as "stop" (exact meaning unclear, but presumably a sudden stop after tracking). The mirror of the telescope was made of K4 glass, and the two correcting lenses of ZK7 and F2 glass, all from Chinese sources. Invar rods maintain the focal distance. A selsyn is used for angular readout.

The essentially finished telescope we were shown is destined for Purple Mountain Observatory, but this is not the first one made; at least one other is already in the field. Tsu and his associates at the factory did not know whether Yünnan Observatory has one. The instrument has an aperture essentially as large as any optical telescope presently available in China for astronomy, although the larger 1-m Zeiss instrument will probably be installed at Yünnan Observatory within about 1 year. In addition to the finished instrument, some of the major steel parts for four or five more such instruments were on hand, so that an excellent network of tracking instruments should be available and in place before long.

FIGURE 5 Satellite tracking camera.

STELLAR ASTROPHYSICS

Research Programs and Instruments

Full-fledged research departments of stellar physics exist at Hsing-lung
Station of Peking Observatory and at Purple Mountain Observatory. There
is also a stellar physics group at Yünnan Observatory, but the staff
there is still small; work in this area will presumably intensify when
the new 1-m reflector (from Zeiss Jena) is installed. Some work in
astrophysics is carried out also at Shanghai Observatory. One of the
divisions of the astronomy department at Nanking University studies the
astrophysics of Sun and stars, but its chief responsibility is to train
students for posts in universities and national observatories.

Peking Observatory: Hsing-lung Station   Hsing-lung Station is at an
altitude of 960 m on a ridge of the Yen Shan Mountains, about 100-km
airline distance northeast of Peking, and has been operating since 1968.
The research programs are best classified in terms of the instruments
employed.
    The 60/90-cm Schmidt is convertible for slit spectroscopy at the *f*/15
Nasmyth focus by insertion of a pseudohyperbolic secondary. The telescope and spectrograph were built in Jena. The spectrograph is of typical
Zeiss design with 75-mm collimated beam, and two plane gratings (ruled

in the DDR) with 651 grooves/mm and blaze angles 10° and 25.°9. There are three photographic field-flattened cameras, the shortest being a semisolid Schmidt of 65-mm focal length. The dispersions available range from 17 to 230 Å mm^{-1}. A typical exposure time that was quoted for this instrument was: a star of B = 10^m requires about 2 hours exposure in the blue at 230 Å mm^{-1} on Kodak 103a-0. The resolution and widening were not specified, but if the spectra are minimally widened this exposure is close to that expected for the given aperture and dispersion. The telescope can also be used as a conventional $f/3$ Schmidt; the corrector is of Schott BK7 glass, and the 5.°3 objective prism of UBK7. With the objective prism, recent work with this telescope has concentrated on a search for Hα emission stars in the Scorpio-Ophiuchus dark nebulas, in coordination with the examination of that area for flare stars with the 40-cm astrograph. With the slit spectrograph, intermediate-dispersion observations of Nova HR Del 1967 and of Nova V1500 Cyg 1975 have been carried out and published. Spectrographic studies of symbiotic stars were also mentioned.

A 40-cm double astrograph (i.e., with two identical optical systems with which duplicate exposures can be made to check for photographic defects, etc.) was also constructed at Jena and was installed at Hsing-lung in 1974. A near or exact duplicate exists at Purple Mountain. Both double astrographs have four-element objective lenses of 3-m focal length. The two objectives on the Hsing-lung astrograph are corrected for a mean wavelength of 4,200 Å. The telescope is used currently in a search for flare stars in dark clouds; a 1-hr exposure on 103a-0 emulsion reaches the eighteenth magnitude. Multiple-exposure plates are also taken in this program with the 60/90-cm Schmidt, consisting usually of 5 15-min exposures per plate; but at the Schmidt a Schott UG2 filter was used because flare amplitudes are larger at shorter wavelengths. This flare-star program has been under way since 1975 and is a cooperative venture with a group at Purple Mountain. Similar intensive searches for flare stars have been or are being carried out at Tonantzintla, Asiago, and Burakan in star fields such as the Pleiades, Orion Nebula, and several dark clouds. It is possible that the interest of the Chinese astronomers was stimulated by the 1959 visit of G. Haro of Tonantzintla, who was then very active in the field. This is an appropriate program for telescopes of this type and aperture when one considers what alternatives exist. We were provided with an unpublished list of new variables in Sco-Oph and in Taurus, which had been found at Hsing-lung (their new discovery numbers are prefixed by the letters "ZB", for "Chinese Variable"). A list of earlier discoveries by this group has just been published. It might be mentioned that, although the number of flare stars (sometimes called flash variables) discovered in young clusters or associations continues to rise, the astrophysical significance of the objects themselves is not at all clear. We saw no evidence that the Chinese astronomers were working on this aspect of the problem.

The 60-cm reflector was built at the Nanking Astronomical Instrument Factory in 1968. It can be used at either the $f/3.82$ prime, the $f/15$ Cassegrain, or the $f/32$ coudé focus. Most of the work is photoelectric photometry carried out at the Cassegrain focus. The practical limiting magnitude of about 11^m is set by the ability of the observer at the guide telescope to see and to hold a star in the photometer diaphragm. (It was with this instrument that the Chinese observations of the occultation of a star by the rings of Uranus were made in 1976.) The main photometric program is the observation of the light curves of eclipsing binaries; an investigation of the yellow light curve of MR Cyg has recently been published.

At the time of our visit to Hsing-lung, a site on a nearby ridge had been cleared for the future 2-m telescope.

Purple Mountain Observatory Purple Mountain Observatory is located on a hilltop at the edge of metropolitan Nanking. The original construction took place in 1929-34, but the observatory suffered extensive damage during the Japanese occupation. Except for the 60-cm reflector, all the instruments that we saw were installed after 1949.

The 40-cm Zeiss double astrograph, installed in 1964, looks very much like the Hsing-lung instrument. It is used part-time in the flare star search program in cooperation with Hsing-lung, and for photographic photometry of novas.

The 60-cm reflector is a pre-World War II Zeiss instrument, installed in 1934. The mirrors were taken to Kunming for safekeeping during the occupation, but the mount was damaged and not returned to service until 1956. The telescope is equipped with an old Zeiss quartz prism spectrograph, producing a dispersion of 150 Å mm^{-1} at 4,340 Å, but its performance is said to be unsatisfactory. Direct photography is possible at the Newtonian ($f/5$) or Cassegrain focus (the limiting magnitude is about 17^m, because of the background from the lights of nearby Nanking). Particular attention has been given to ultrashort ($p \leq 0^d.1$), hot (B-V ≈ 0.0) cepheid variables in globular clusters and in the field. The spectroscopic work at the 60-cm telescope has been concentrated on novas, Be, and symbiotic stars. We heard a report on photographic spectrophotometry of X Persei, the type Be candidate for identification with the X-ray source 3U0352+30, which demonstrated that the interstellar extinctions of the two objects are essentially the same, thus supporting the identification. The Purple Mountain staff intends to upgrade the performance of the 60-cm reflector, but their ability to produce a competitive instrument will depend upon the availability of photographic emulsions, gratings, detectors, etc. from abroad.

Shanghai Observatory: Zo-se Station It is appropriate to mention here the work on proper motions of clusters being carried out at Zo-se Station of Shanghai Observatory, although we were unable to visit Zo-se because of reconstruction work in progress there. The telescope is a double astrometric refractor of aperture 40 cm, focal length 7 m, dating from the time of the former Jesuit college (1900-1950). The Zo-se archives contain an especially large collection of plates of the Orion nebula region obtained in 1902-16, which were repeated in 1975 and have now been used for a study of the proper motions of emission and OB stars in that association. The OB stars indicate a small group rotation and expansion. (The cluster M67 has also been studied astrometrically; the plate material for M8 has not yet been worked up.)

Future Development

It is obvious that the Chinese astronomers, although energetic and very well informed (their familiarity with the Western literature was apparent from the keen questions they asked), are severely limited in their ob-servational work by the telescopes and instrumentation now available to them. It is equally clear that they intend to do a good job with what they have. We were questioned at length and in detail on all kinds of instrumental matters, such as photographic materials, hypersensitization, calibration procedures, and attainable accuracies of all types. The designers at the Nanking Astronomical Instrument Factory, on the other hand, were looking ahead to the instrumentation to be provided for the 2-m reflector. A large spectrograph and a Fourier-transform spectrometer are planned for the coudé focus, and another spectrograph whose nature is not yet decided will go at the Cassegrain. No consideration has yet been given to prime focus equipment. Again, from the questions asked it is obvious that the designers are still turning over in their minds the various advantages and trade-offs.

RADIO ASTRONOMY

Telescopes in Operation

Each of the major observatories in China has some radio astronomy instru-mentation. Sha-ho Station of Peking Observatory, Yünnan Observatory, and Purple Mountain Observatory all have similar instrumentation for solar observations. They use small, equatorially mounted reflectors of 1.5- to 3-m diameter, together with simple radiometers operating at 3.2 cm (all three observatories) and 10 cm (Sha-ho and Purple Mountain, although Yünnan is planning to install 10-cm equipment in the future). In all of these installations the receivers are Dicke-type switched radiometers, switching against a room temperature load, with uncooled mixer first stages and Klystron or Gunn local oscillators. They have system temper-atures of about 2,000 K, and, while rather primitive by today's standards, they are quite adequate to monitor the total flux of the Sun at 3- and 10-cm wavelengths.

A more sophisticated instrument for solar observing is in operation at Mi-yun Station of Peking Observatory (Figure 6). It consists of an east-west array of 20 parabolic antennas, each 9 m in diameter, operating at 450 MHz as a compound interferometer. With a total length of about 2,300 m, the array provides one-dimensional resolution of about 1 arc min. Adjacent lobes are spaced 67' apart, so only one beam is on the Sun at a time. An open-wire transmission line is used, feeding four transistor preamplifiers. The system bandwidth is 1.5 MHz, and the noise temperature about 500 K.

The 450-MHz interferometer has been in use since 1976. It evolved from an earlier system, built in 1967, consisting of 16 6-m diameter reflectors operating at 146 MHz. In 1975 the reflectors' diameters were increased to 9 m and the 450-MHz electronics added. The 146-MHz system is no longer used. It is still in existence, however, and parts of it will be used in a new aperture-synthesis array, described below.

FIGURE 6   Radio antenna array at Mi-yun Station.

The 450-MHz interferometer gives a good signal/noise ratio for the Sun and provides marginally useful resolution for monitoring active regions. While we were in China, severe interference of unknown origin was being experienced. It was strong enough to saturate the receivers and make observing impossible. This is a common problem throughout the world. We hope that the Chinese will be able to determine the nature and origin of the interference and take steps to eliminate or minimize it.

The 9-m diameter antennas are equatorially mounted. They were fabricated at the site by the observatory staff, and are said to be adequate for work at 21 cm.

The astronomy groups at Nanking and Peking universities each have small radio telescopes for teaching purposes. At Peking University there are an 8-m diameter reflector and a 21-cm receiver for solar observing, while Nanking University has a 2-m dish operating at 3.2-cm wavelength, again for solar observing.

The instruments noted above are the only operating radio telescopes we saw. In addition, radio astronomy work has recently been initiated at Shensi Observatory, which we did not visit. There is a small dish for solar observing at 3.2 cm. Further plans at Shensi were not described.

Present Construction and Future Development

Development work or planning for future instrumentation is under way at all of the observatories we visited. Currently under construction at Mi-yun are two new instruments. Twelve additional 9-m diameter reflectors are being built, each of which will be connected to each of 16 of the existing antennas to yield an aperture synthesis array with 192 east-west baselines, at 6-m intervals from 12 to 1200 m. The antennas will have electrical drive in hour angle, and declination. The array will operate at 176 MHz, using the old 146-MHz transmission line system for local oscillator distribution. There will be a 176-MHz paramp at each antenna. Since only a 96 correlator back-end will be available, it will be necessary to observe a given region twice, each time with 6 × 16 baselines, to do a full synthesis. Correlator outputs will go on an on-line minicomputer for processing. The specific computer has not yet been selected. A new, modern control building has been completed to house the array controls, minicomputer, electronics, etc. The array should be completed in about 2 years. It will have a 10° field of view and give resolution of 3.8 arc min. and sensitivity of 0.15 Jy in a 12-hour observation. This instrument, if completed on schedule, should rank among the major radio astronomical instruments in the world. Its sensitivity will be comparable to that being obtained anywhere at these frequencies.

The aperture synthesis array shares 16 of its antennas with the 450-MHz interferometer described earlier. The intent is to do solar observing in the daytime at 450 MHz and aperture synthesis observing at nighttime at 176 MHz.

A second aperture synthesis array is also under construction at Mi-yun. It will consist of 20 dishes, 1.5 m in diameter, operating at 3-cm wavelength. The array will be on an east-west line about 100 m long. It will make use of (in the daytime) the same 30-MHz correlator system and minicomputer used at nighttime for the 176-MHz array.

At Shanghai Observatory there is interest in developing VLBI techniques for use in the Chinese time service. They have set up a 6-m diameter antenna for experimental work, but do not yet have any specific plans.

At Purple Mountain Observatory the principal interest for the future appeared to be in millimeter-wavelength astronomy, and discussions are under way concerning the direction they might go in that area. They are now building an 8-mm receiver and a 60-mm reflector for solar observations, as part of their millimeter-wavelength development. The 8-mm receiver consists of a ferrite switch for Dicke switching between load and antenna, a transistor paramp, and a Gunn local oscillator. It has a 70-MHz bandwidth and about a 3,500-K noise temperature. It was being laboratory-tested while we were there. The 8-mm waveguide components were commercially made in China and appeared to be of excellent quality.

EXTRAGALACTIC ASTRONOMY

In the People's Republic of China a great deal of attention is devoted to theoretical cosmology and extragalactic astronomy, principally at Peking Observatory, Peking University, Purple Mountain Observatory,

Nanking University's astronomy department, and the Chinese University of Science and Technology. Virtually no observational work is done in these fields, but a thorough knowledge of Western literature has resulted in the interpretation and analysis of observations made by others.

Galaxies

The interest of Chinese astronomers in extragalactic research has been stimulated by two sources. First, astronomers in China have ready access to current literature and journals, although most of the journals are photocopies with relatively poor reproductions of the half-tone plates of galaxies. The worldwide interest in galaxies emphasized in these journals is clearly shared by the Chinese.

Second, Chinese interest in extragalactic research has been kindled by the visits of Lin Chia-chiao (C. C. Lin of M.I.T.) to Peking and elsewhere during the past year and a half. C. C. Lin, as an "Overseas Chinese" who has attained academic distinction in the United States, has clearly had a very strong impact on the very intelligent and able theoreticians whom we met, especially at the Peking astronomical centers. C. C. Lin and his collaborators and students developed an elegant mathematical theory of "density waves," which provides a long-sought solution to the puzzle of the persistence of spiral structure in flattened disk galaxies and the limited degree to which the spiral arms wind up, despite the number of revolutions of the order of 100 with strong differential rotation during the lifetime of a typical galaxy. The attraction of this theory from the mathematical point of view has appealed to those Chinese astronomers who are mathematically and theoretically gifted, and the fact of C. C. Lin's Chinese origin has created a particular interest in his work.

During the symposia in Peking and in Nanking we heard many reports on various aspects of density wave theory, mainly concerning questions of stability and the problems of maintaining a spiral pattern in galaxies with disks of finite thickness.

Institute of Geophysics In Peking, at the Institute of Geophysics of the Chinese Academy of Sciences (CAS) which includes the fields of astronomy and astrophysics, we spent two days (October 6 and 7) listening to papers by the Chinese, interspersed with talks by Heeschen, Sandage, and Burbidge, all on extragalactic topics. In one of the Chinese talks that dealt with galaxies, Liu Yong Chen presented a modification of C. C. Lin's treatment by adding source and sink terms to Lin's equation to allow for star formation in the density waves and the consequent heating. In this way he could reproduce some observed features of the spiral arms in galaxies.

Li Ch'i-pin showed that radial expansion of galaxies does not affect the persistence of density waves and hence of the spiral arms. Chou Cheng-long (with Ch'en Chien-sheng) studied the effect on the Jeans instability of combining gas and stars by assigning different velocity dispersions to the two components.

The session on October 6, our first encounter with large numbers of Chinese scientists, took place in a lecture hall containing perhaps 150 people. We learned later that the audience consisted of students, faculty members, and research workers at several institutes of the CAS in the Peking area, as well as visitors from other universities who had traveled to Peking for these meetings.

We arrived very late, but despite our considerable delay the audience was sitting packed in the lecture room, patiently waiting for us, and greeted us enthusiastically. They did not have many questions for the three American speakers; it was our impression that they have had little exposure to presentation of observational results in lecture form. Our illustrations were given in the form of slides, which is clearly not common practice; in their own papers, the Chinese astronomers used clips of blackboard-sized sheets of paper on which they had written their equations, etc., and which they hung on a blackboard, turning over the leaves in the way we put sheets on a View-Graph.

The relative formality of the October 6 and 7 sessions at the Institute was in strong contrast to the free and easy exchange that took place at Hsing-lung Station, where four of us stayed the night of October 4 in a new dormitory. As we sat in the recreation room at Hsing-lung during the evening, when observing was impossible because of the rain, we invited our hosts to ask us anything they wanted. This opened the floodgates, and they rapidly lost their inhibitions; despite the language problem, communication was immediately easier. They asked detailed questions on the spectra of QSO's; on how one determines their redshifts when only a few emission lines are visible; about the determination and form of velocity curves (velocity plotted versus distance from the nucleus) in galaxies; and about the active nuclei of galaxies.

Our entire evening's discussion at Hsing-lung was cordial and relaxed. At Li Ming-te's insistence, we finally adjourned the session and retired to bed at 1:00 a.m.

Our last formal scientific sessions in Peking took place at the university on October 8. After the greetings and tour, we heard further papers by the Chinese astronomers. In contrast to the almost entirely cosmological, theoretical, extragalactic sessions at the institute, these lectures were varied. The only paper on galaxies was P'eng Ch'iu-ho's on "Mass Distribution in a Disk Galaxy of Finite Thickness." He used cylindrical rather than spheroidal coordinates (this involved some mathematical complexity that spheroidal distributions avoid) and compared his results with Toomre's thin disk model.

The final session with the Peking astronomers took place on Sunday, October 9, at the Peking Hotel, where we divided into two groups for informal discussions. At the group discussion on extragalactic topics, the problem of the redshifts was raised: that is, are there any noncosmological redshifts in QSO's? Burbidge and Sandage staged a "controversial discussion" to demonstrate how free argument and give-and-take goes on between friends in the Western scientific scene.

Several Chinese then made quite penetrating comments on the "big bang" theory of cosmology. There was a good discussion about the early stages of the universe, with emphasis that the true picture is undoubtedly more complex than the treatment in current studies in the West. Of special concern is the problem of the present nonhomogeneity of the universe versus the isotropy of the cosmic background radiation. Questions were also asked about the spatial distribution of the QSO's and about whether the apparent limit to their redshifts (z = 3.5) is real or due to observational selection.

From this discussion it is clear that, when the Chinese have a telescope of modern size and capability (i.e., the forthcoming 2-m), they will enter directly into extragalactic observational programs and that they are gathering plans and ideas in preparation for this. However, we noted that there are clearly a handful of leaders in this field; a majority of the group (perhaps partly because of the language problem and residual shyness) did not join in the discussion.

Yünnan Observatory Our next stop after Peking was Kunming. Because of lack of space at the observatory, the scientific sessions were held in the Kunming Hotel. Our hosts had requested talks from us on a long list of topics, and it was clear that only by dividing into two groups could we come close to covering their requests in the time available.

Burbidge, Heeschen, Herbig, and Sandage took the session on QSO's, redshifts, and detectors for faint objects. The only paper on galaxies presented by the Chinese at this session was Huang Yin-liang's "The Asymptotic Solution of Poisson's Equation for a Screw-shaped Disturbance of the Density of a Spiral Galaxy of Finite Thickness." This astronomer is clearly a very gifted theoretician. He set out (with large sheets of paper, as in Peking) the three-dimensional Poisson's equation in cylindrical coordinates, and then proceeded to simplify the solution by taking a two-armed spiral and putting in sizes appropriate for our galaxy.

Our own talks elicited a large number of questions; those on QSO's were still going strong at the end of our session, at 5:30 p.m.

While not so much extragalactic work is currently going on at Kunming, there is clearly great interest in this field. Since Kunming will shortly have a 1-m telescope, it is likely that some observational programs will be started (e.g., wide- and narrow-band photometry of galaxies could be done).

Shanghai Observatory Our next stop, Shanghai Observatory, revealed a very high level of sophistication. Shanghai has been a center of astronomy for decades. Our stay here was very brief, and we had one morning scheduled for talks by the Chinese astronomers. The talks on galaxies were much more related to observational aspects than had been the case in Peking. Wan Lai presented "Correlation Analysis of X-ray Luminosity and Cluster Velocity Dispersion," a topic of current interest in extragalactic research in the West. Up-to-date Western observations were used (Faber's from Lick Observatory). Tai Wen-sai (just up from a hospital bed; he had come to Shanghai from his base in Nanking for a serious operation) discussed the Hubble classification of galaxies and whether it represents a cosmogonical sequence or is a result of initial conditions. His paper gave an impressive analysis of the observational data on angular momentum of galaxies. He showed a new result that the "specific angular momentum," i.e., $J/M^{7/4}$, where M is the mass of a galaxy with angular momentum J, is constant along the Hubble sequence. The result, not understood, is clearly fundamental and is also related to the question of what ranges of mass and angular momenta lead to the formation of double and multiple galaxies rather than single systems.

Chao Chün-liang spoke on "The Distribution of Galaxy Clusters and Radio Sources in Space," and, here again, his work relied heavily on up-to-date Western observational data--in this case, mainly radio-astronomical results from Greenbank. He discussed his calculations on the distribution of clusters of galaxies and of radio sources by binning analysis, along the earlier lines of Limber (*Astrophysical Journal 119* p. 655, 1954) and more recently of Peebles and his collaborators, showing non-random correlations on all cell sizes.

Purple Mountain Observatory We also found the work at Purple Mountain Observatory to be more closely related to observations than at Peking, but the observations tend to be of a type current some decades ago in the United States. Astronomers there are particularly interested in stellar evolution, and they asked for talks on this and on galactic structure and also on radio galaxies and galaxies with active nuclei. Lu T'an presented a paper, "Anti-matter Model of Quasars and Active Galactic Nuclei," a theory that has been advanced by Alfvén and co-workers in the West but that has run into problems on the flux of γ-rays that would be expected and also on the cosmological model that can accommodate anti-matter in equal amounts with matter in the early stages.

Quasars (Quasi-Stellar Objects, QSOs)

A great deal of interpretive work on the astronomical aspects of quasars (distributions, linear sizes, use of the Hubble diagram for cosmology, separation rates that apparently exceed the velocity of light, etc.) is being done, mainly by a group at the Chinese University of Science and Technology. Many of the members of this group came to Peking and again to Nanking to contribute to the symposia.

In both Peking and Nanking members of this group gave papers on various quasar parameters that could be calibrated as "standard candles." By analyzing data in Western literature, the authors found that the largest distance between the radio components of quasars with double radio structure appears to be correlated with the optical luminosity of the quasars. Thus this appears to be a useful parameter. Assuming cosmological redshifts, they plotted a redshift-distance relation that they found to depart from the local Hubble relation for large z, which is to be expected if it is due to deceleration. A formal value of $q_0 \simeq$ + 1.4 would explain the result. This group did indeed take the normal view as part of their analysis that the Hubble law is, in fact, linear locally everywhere. Their conventional view in this matter is in contrast to others at the Peking symposium (see the next section) who argued that the expansion is nonlinear ($V \propto r^2$) a point on which a great deal of discussion occurred between the Chinese astronomers and our group.

Another talk on the radio properties of QSOs was given at Purple Mountain Observatory on October 18. Three of the authors of the other quasar work described above derived an average and an upper limit for the velocity of separation of the radio components by analyzing the observed radio flux density ratio of the two radio components for 26 QSOs with double radio components. They found $v_{max} = 0.22c$ and the average

$\bar{v} = 0.09c$. In their analysis they used a variety of Western source material, all quite current.

However, some of the most recent Western observational work on QSOs
has either not reached the Chinese astronomers or has not yet been in-
corporated into their theoretical work. In particular, they do not seem
to have pursued the physical models of line-emission regions in QSOs,
perhaps because of the lack of a strong past tradition in astronomical
spectroscopy. They have not devoted thought to the problem of the
multiple-redshift absorption-line systems in the spectra of some QSOs
and had no questions on this topic following the brief account incorpo-
rated in one of our talks in Peking.

They have, however, devoted some thought to the evolutionary connection
between QSOs and galaxies and are deeply interested in the fact that
the QSOs have very large redshifts. Many accept that these redshifts
are cosmological (i.e., due to the expansion of the universe) while a
few leave open the possibility that there might be a non-cosmological
component. They ask about gravitational redshifts, and, like Western
theoreticians, have no alternative theory for redshifts other than
Doppler.

Cosmology

The work in cosmology that we saw can be divided into three general
fields: (1) basic theory of the Einstein equations of gravitation as
they pertain to the global properties of the universe; (2) analysis of
data on redshifts, radio fluxes, angular separation of double radio
sources, and spatial distributions of galaxies, clusters of galaxies,
and quasars as related to world models; and (3) questions of the formation and evolution of galaxies described earlier.

We heard two papers on the foundations of cosmology by members of
a Peking group that works in three CAS Institutes (Institute of High
Energy Physics, Institute of Mathematics, and Institute of Physics).
The papers dealt with (1) a generalization of the Einstein formulation
of the gravitational-field equations in terms of gauge symmetries of
certain transform groups (the Lorentz, Poincaré, and de Sitter groups),
and (2) an application of the result of this attempted generalization
of the Einstein theory to global predictions with observational consequences (redshift-magnitude relations, etc.). The unconventional predictions of this work are: (1) claims for a removal of the necessity
of collapse to a black hole, and (2) a nonlinear velocity-distance
relation ($V \propto r^2$). A lively discussion resulted, both in the formal
session and later informally, concerning the observational evidence
against the ($V \propto r^2$) prediction. It was clear that the theoreticians
carrying out this work did not have full information on the current
observational evidence on the linearity of the Hubble expansion, or on
the effect of observational bias (the "Malmquist effect") on the contrary
claims of Hawkins and Segal, whose work the Chinese astronomers quoted.

A most interesting subsequent discussion was held with a group of
about 40 theoreticians, including the authors of the paper on a ($V \propto r^2$)
relation. It appeared that Western conventional views of Friedman-Hubble
cosmology are, in fact, held by a large number of Chinese astronomers.

Other papers at the October 6 and 7 symposia concerned practical aspects of cosmology. Sun K'ai of Peking University described his work on the luminosity function of nearby field galaxies, giving an elegant formal solution to this classical problem, with applications to the data on redshifts and magnitudes that are presently available in the literature. He discussed the effects of catalog incompleteness, and will surely carry the work further when more complete redshift data for nearby galaxies are available.

As mentioned earlier in the section on quasars, much work on the properties of quasars is directed to using them as cosmological probes in order to find ways to reduce the effects of the large dispersion in absolute magnitude. In this connection, studies of the Hubble diagram for quasars were discussed, and results related to q_0 were shown. As a generalization and extension of the discussion, we exchanged views on the current ways to determine q_0 using many different methods, including the deuterium problem and the time scales of stellar evolution compared with the Hubble constant. The first-rate discussion was led by Fang Li-chih, one of the members of the group from the Chinese University of Science and Technology.

Work in high-energy astrophysics was described principally at the Nanking symposium. Papers on "Neutron Stars," "The Spin-Down Rate of Pulsars," "An Anti-Matter Model of Quasars," and "Active Nuclei of Galaxies" were given, together with papers on "The Nuclear Physics $(e^+, e^-$ pairs) of the Accretion Disk Around the Black Hole of Cyg X-1" in an effort to model the observed hard (100 keV) X-rays that no previous model had succeeded in explaining, and the propagation of low-frequency ($\nu < 30$ kHz) waves in a relativistic gas in a magnetic field, with applications to energy transport in the Crab Nebula.

The general impression from this extensive work in extragalactic astronomy is that the level of theoretical activity is high and of excellent quality. The potential is clearly present to make original, significant contributions, provided that direct contact can be established with observational astronomers who will be working in China when the 2-m telescope is completed. If the contact can be maintained and strengthened as the new instrumental equipment becomes competitive, there is every reason to believe that Chinese work in cosmology can attain worldwide significance. The key clearly now lies primarily in acquiring the crucial observational equipment.

ELECTRONIC COMPUTERS

Electronic computers are playing an increasingly decisive role in modern astronomy, a fact well recognized by Chinese astronomers. We had the opportunity to visit two computer installations, one relatively large and one much smaller.

The large computer is located at the Institute for Computer Technology at Peking, one of the institutes of the Chinese Academy of Sciences. Its technical characteristics are: a word length of 48 bits, a speed such that a multiplication takes about 3 µs; a main core memory with a capacity of 130,000 words; and a major magnetic disk memory consisting of 2 units each containing 10 disks, as well as 8 magnetic tape units using a 16-channel code (8 data channels, 5 check channels, and 3 synchronization channels).

The input equipment consists of 4 paper-tape readers with a speed of 800 bytes per second, a control typewriter, and a CRT character display unit with a keyboard. The main output equipment consists of 4 line-printers with line lengths of 120 characters with speeds of between 360 to 600 lines per minute. However, surprisingly, no card readers or card punchers are included in the input and output equipment. The machine was assembled in Peking from components produced entirely in China.

The computer is used mainly by computer and systems designers, in accordance with the principal goals of the institute in which it is housed. However, ample time is also available to engineers, medical specialists, and other scientists, among which astronomers have been from the very beginning persistently active users. The computer operates 24 hours a day with 1 to 2 hours used for routine maintenance. The scientists generally use Fortran IV or a language designated as "BCY", which was described as similar to Algol 60. The scientists write their own codes, but the punching of the paper tapes is done by special workers. The use of this computer, probably the biggest one presently available to Chinese astronomers, is so well organized that the turn-around time for a moderate sized astronomical computation is frequently as short as 1 hour.

At present the computer is not connected with terminals outside the Institute for Computer Sciences. However, separate paper tape punches (for the preparation of input tapes) exist in several places. This relatively large computer is not used for training students for which purpose Peking University has some small machines.

The smaller electronic computer we visited had just been installed in temporary quarters at Kunming Observatory. The machine was still in the testing stage so that the overall efficiency of its use had not yet been assessed. Its designation is TQ-16 and its technical characteristics are: a word length of 48 bits, a core memory of 32,000 words, and an operating speed about 10 times slower than that of the bigger machine just described. Two magnetic drum units and 2 tape units augment its memory. Its main input again operates through paper tapes (not punched cards) and its main output unit consists of a printer, designated JY-80, which has a line length of 80 characters and a speed of 600 lines per minute. Altogether this smaller computer appears to be roughly comparable with the CDC 1604, which went into production about 1960.

The TQ-16 computer of Kunming Observatory may be expected to be entirely dedicated to astronomical work, presumably including the reduction of satellite-tracking data and the computational solution of the corresponding celestial mechanics problems. Another TQ-16 computer, located in the Nanking Astronomical Instruments Factory, is used for lens design. Furthermore, we understand that Purple Mountain Observatory will soon install a TQ-16 computer, mainly for the purpose of orbit calculations.

HISTORY OF ASTRONOMY

Sivin's meetings with historians of science in Peking, Nanking, and Shanghai, began in each case with an exchange of lectures (see Appendix C). In addition to reporting on his own work, Sivin summarized research under way on the history of Chinese science in Western Europe, Japan, and the United States. The more senior of the Chinese historians were generally familiar with the published literature and were prepared to ask Sivin detailed questions about his own books. His summary of research by historians in the United States, Western Europe, and Japan was therefore devoted mainly to work in press and underway.

Following the exchange of lectures, the meetings proceeded to informal discussions of current research trends. The possibility of increased international communication among scholars in the field was also explored. In the course of these informal conversations, the Chinese provided information on individual and collective research projects under way and planned. A brief summary of research trends follows below. A detailed report has been published separately by Sivin in *Chinese Science*, No. 3, 1978.

In China research on early mathematics and astronomy has been predominant in the historical study of science since publication by people trained in modern science began in the second decade of this century. (Contributions to the history of science by scholars trained primarily in history are rare in China and are generally a recent phenomenon in Europe and the United States.) Largely because most of the leading figures such as Li Yen, Ch'ien Pao-tsung, and Yen Tun-chieh were interested in mathematics and secondarily in astronomy, the trend toward specialization in these two fields continued until the Cultural Revolution.

The Research Institute for the History of Science was organized within the Chinese Academy of Sciences in 1954. Its *Journal of the History of Science (K'o-hsueh-shih chi-k'an)* reported the results of research on foreign, as well as Chinese, science. The period between 1955 and 1966 was one of exceptionally fertile scholarly publication. In addition to monographic works in every field of the history of science and technology, well-documented general histories of early Chinese mathematics, astronomy, chemistry, and mechanical technology appeared; bibliographies of mathematics, astronomy, agriculture, geology, military technology, and other fields, as well as other reference works, were compiled; and large numbers of important primary sources were republished, both in facsimiles of rare editions and in modern editions. This activity continued until the Great Proletarian Cultural Revolution began in 1966, after which the Research Institute of the History of Science was closed and its journal ceased publication.

When a few scientific journals resumed publication in 1973, they provided space for accounts of the research that had gone on in the meantime. For instance, the lead article in the first issue of the new series of *K'o-hsueh t'ung pao*, a general science journal, commemorated the Copernicus Quinquecentennial. Essays on the history of astronomy have appeared frequently in the major Chinese astronomical journal and in other scientific periodicals. Papers that bear on early astronomy have been published frequently in journals of archeology. Whether the *Journal of the History of Science* will reappear depends, of course, on the outcome of policy discussions that seem to be in progress.

It appears that a burst of activity in the historical study of Chinese science similar to that of 1955-66 is in the making. Institutions to support research are once more visible to foreigners. With the end of the Cultural Revolution, a number of research institutes, some of which had been closed, were combined to constitute the new Chinese Academy of Social Sciences. A new research institute for the history of natural sciences has been functioning since January 1978 under the Chinese Academy of Sciences. Shanghai and Purple Mountain observatories include research groups for the history of ancient astronomy (these observatories are subordinate to the Chinese Academy of Sciences). These research groups appear to be in regular touch with the larger research institute in Peking, but at the same time their work is closely integrated with that of their astronomical colleagues, and a number of members do both astronomical and historical work. Chang Yü-che, the distinguished director of Purple Mountain Observatory, for instance, is an enthusiastic part-time participant in the work of the Research Group on Ancient Astronomy at his own institution; he recently finished a research project on the orbit of Halley's comet over the past 4,000 years.

A history of Chinese cosmological conceptions from earliest times to the nineteenth century, by Cheng Wen-kuang and Hsi Tse-tsung, was published in 1975. General histories of astronomy and mathematics, a volume of pictorial documentation for the history of astronomy, and a general study of traditional Chinese science from the sixth century on are among the projects now under way.

Since work in the history of science began to appear in print again in 1973, the role of astronomy has become even more important than before. This is due in part to the continued intellectual leadership (alongside that of Yen) of Hsi Tse-tsung, whose main contributions have been in the history of astronomy. A second reason is that, during part of the last decade in which many research institutes ceased to function, astronomical observatories for the most part remained at work. They were able to provide occupations for some scholars trained in astronomy and experienced in historical research and to provide a working environment in which astronomers could explore the historical record for data of current utility. A third reason has to do with the emphasis given to archeology by national research policy since the 1950's. Earlier in this century the study of artifacts by historians of science was not infrequent, but it was considered largely a supplement to study of the written record. The continued strength of archeological fieldwork and research during the Cultural Revolution and its aftermath provided many opportunities for collaboration, which historians of science seized. Since 1950 work in museums and laboratories has come to have at least as high a status as library research. Since 1966 a great deal of the significant publication in the history of astronomy, and most of that in the other fields of the history of science and technology, have related to archeological discoveries, discoveries unprecedented for any similar period in both quality and diversity. The fields that have

benefited most are technology and astronomy. A few important finds
bear on ancient biology, mathematics, and so on, but current understanding of astronomy in early China has been strongly affected by a stream
of remarkable finds, the most significant of which are:

1. Many inscribed bones and tortoise shells from the fourteenth
to eleventh centuries B.C. (Honan, 1973);
2. An astrological manuscript of the second century B.C. that includes a planetary ephemeris for 246-177 B.C., information on planetary
periods, and illustrations of comets and other celestial phenomena
(Hunan, 1973);
3. An ephemeris for 134 B.C. (Shantung, 1972);
4. The platform of a central government observatory build in A.D.
56 (Honan, 1974-75);
5. A portable folding bronze gnomon and shadow template used for
solstice and noon determinations, first or second century A.D. (Kiangsu,
1965, correct identification not yet published);
6. Three bronze water clocks of the Han period (second century B.C.
to second century A.D.; from Hopei, Shensi, and Inner Mongolia, the
third not yet published);
7. A great variety of star maps, mostly painted and carved in tombs,
from the second or first century B.C. to one of A.D. 1116 which depicts
the signs of the Western zodiac.

Topics of current research tend to be divided between work on recently
discovered artifacts and manuscripts, largely in collaboration with ar-
cheologists, and study of ancient documents to yield data useful in the
solution of contemporary astronomical problems, a number of which have
been mentioned in Chapter 1. The research reported on either to Sivin or
in lectures to the delegation at large involved a broad spectrum of meth-
odologies for the understanding of ancient technical accomplishments.
To give only three examples, a report of planets and comets illustrated
in a manuscript of the second century B.C. applied a modern classification of comet types to find a pattern in the diversity of comets (over
20 types) shown; a study of lunar eclipses recorded on inscribed bones
of *circa* 1300 B.C. ingeniously applied linguistic and chronological analysis to cast light on the early calendar; a third, discussed above,
applied autocorrelation analysis to a large number of Chinese sunspot
records made over the last 2,000 years to detect cyclic behavior. (A
complete list of historical lectures is included in Appendix C.)

In view of the variety of research topics, the revivification of
institutions to support research, and the new circumstances that make it
easier for Chinese historians of science to be in touch with their colleagues the world over, it is likely that the history of astronomy in
China will soon be producing scholarship equal in quantity, as well as
quality, to that published in the 1950's and early 1960's, and will ultimately surpass earlier accomplishments.

6

INSTRUMENTS AND FACILITIES UNDER CONSTRUCTION

To the American delegation, Chinese astronomers appeared to be very aware of the shortcomings of their research instruments. For about a decade, starting in 1967, little or no research instrumental development took place. Now that restrictions for such development have been lifted, the astronomers are moving fast to modernize their observational facilities. They have been striving to implement the policies adopted by the Eleventh National Congress in order to catch up as quickly as possible.

OPTICAL INSTRUMENT DEVELOPMENT

2-m Telescope

The foremost projected instrumental development project is the construction of a 2-m telescope at the Nanking Astronomical Instruments Factory. This project is being carried out with the cooperation of the astronomers of Peking Observatory. To be located at Hsing-lung Station, this telescope, when completed, will be China's principal optical astrophysical research instrument and will be available to all Chinese astronomers. Initially only the Cassegrain and coudé foci will be available.

The primary mirror blank was cast in Russia and acquired by China about 15 years ago. The glass blank has three fused layers of a pyrex-like material designated as ZK7. The blank has a diameter of 2.16 m and is 32 m thick. When shown to the American delegation, the blank was mounted on a vertical lathe (Figure 7) being used to drill 18 holes in the rear surface where the back-support mechanical cantilever units will be inserted. A central 550-mm opening is also to be drilled, while the 3,000-kg blank is mounted on the vertical lathe.

After these preparations the blank will be transferred to a grinding-polishing-figuring machine built in China prior to 1967 (Figure 8). On this machine the blank will rest on a special cell, which is not the one that will support the mirror at the telescope.

Present optical design specifications call for an $f/3.0$ primary with an $f/9$ Ritchey-Chretien focus in addition to an $f/45$ Coudé focus. The optical configuration will be a scaled-up version of the one

FIGURE 7 Mirror blank
for the new 2-m tele-
scope mounted on a ver-
tical lathe.

described in *Acta Astronomica Sinica* (Vol. 17, No. 1, 1976). A two-
element Ritchey-Chretien focus corrector will yield images less than
0.3 arc-sec in diameter over a 0°.9 (300 mm) field for direct photography with 30- × 30-cm plates.

The optical configuration for obtaining a coudé focus is innovative in
that no change to a special coudé secondary mirror is required. Instead,
a 45° flat mirror centered at the intersection of the primary mirror's
optical axis and the declination axis will direct the beam to another
45° flat mirror inside the polar axis of an English-type mounting. The
latter mirror, being close to the Ritchey-Chretian focus, will be small.
It will direct the beam towards a 36-cm oblate spheroidal mirror at the
north end of the polar axis. The latter mirror will transfer the image
from the Ritchey-Chretien focus to a conjugate focus at the slit of the
coudé spectrograph. Thus, to change from normal Ritchey-Chretien optics
to the coudé configuration will require only inserting the No. 3 flat
mirror in the proper position. Excepting the primary mirror, all the
other mirrors will be made of CERVIT. The Ritchey-Chretien focus camera
correctors will be made of fused quartz.

Present specifications for the optical figuring call for 0.1λ surface accuracy and 90 percent of the light to be concentrated within 1
arc sec of the stellar image. The opticians hope in fact to do better,
and in discussions at the Nanking Astronomical Instrument Factory we
stressed the desirability of just that. We pointed out to them that
the CTIO 4-m primary mirror concentrates 100 percent of the light

FIGURE 8 The grinding-polishing-figuring machine to be used in preparing the primary mirror for the new 2-m telescope.

within a 0.6" diameter disc, 87 percent within 0.3", and 60 percent within 0.1", so that such performance is attainable. We also pointed out that the optics should be good enough to take advantage of the best seeing to be encountered, not just the average. We did not know just how good is the seeing at Hsing-lung. We had been told that the best seeing at the zenith on the Danjon scale (i.e., estimated visually from the appearance of diffraction rings) was 0.25", but this is not readily converted to image diameter, nor had we heard how often such seeing occurs. The Nanking engineers acknowledged the importance of these considerations.

Plans for the mounting call for a Serurrier-truss tube, as shown in Figure 9. Radial support of the primary mirror will be provided by cantilever units. The north and south bearings of the polar axis will be oil-pad type. Thought is being given to using a single D.C. torque motor driving an encoded crown-worm gear combination for moving the telescope in right ascension as well as in declination. During our visit we discussed the possibility of using spur gears rather than worm gears. It is possible that the former type of gearing may be used.

We were impressed by the thought and competence that had been applied to the optical design of the 2-m telescope in both the $f/9$ Ritchey-Chretien system and in the quite original $f/45$ coudé arrangement. As the result of a series of questions from both sides, our

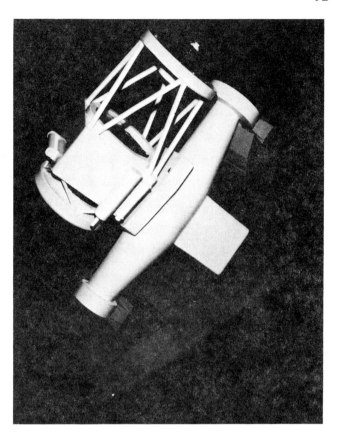

FIGURE 9 Model of the
new 2-m telescope.

understanding of the designers' thinking was clarified. We took the
opportunity to stress some considerations of operation, efficiency,
and convenience that had come from our experience with large telescopes in the United States.

The auxiliary instruments now being considered for the 2-m telescope are:

- Ritchey-Chretien camera with correctors
- Ritchey-Chretien photometer
- Classical Cassegrain spectrograph
- Fourier transform spectrograph for the Coudé focus

At the time of our visit only preliminary planning for these auxiliary
instruments had been carried out.

1-m Telescope

While planning and initial work are progressing on the 2-m telescope,
a 1-m Cassegrain telescope is being acquired from Zeiss Jena. The

telescope will be installed at Yünnan Observatory. It is expected that this telescope will be operating during 1978. Like the 2-m telescope, this one will be made available to all Chinese astronomers.

The 1-m Zeiss telescope will have $f/13$ Cassegrain and $f/45$ coudé foci. At both of these foci, spectrographs ordered from Zeiss Jena will be operated. A Cassegrain direct camera will be available, as will a Chinese-made photometer.

The optical specifications required from Zeiss Jena for this telescope will call for 90 percent light concentration within 0.7 arc sec. In view of more favorable concentrations achieved with the large reflectors now operating at CTIO, KPNO, ESO, and AAO, it is hoped that the telescope will be delivered with better-quality optics than called for in the specifications.

Future Expansion

The Chinese astronomers realize that larger telescopes than the 2-m will be required to put them at the forefront of astronomical research. Experience in the design, construction, and operation of such telescopes must first be acquired, however. Thus the 2- and the 1-meter telescopes will play essential roles in the near future, for they will provide the necessary experience. Beyond putting these two instruments in operation, they are not now actively planning for any additional telescopes. Nevertheless, leading Chinese astronomers are considering what should be the next major instrumental development after the 2-m telescope is operating. In informal discussions, delegation members expressed the opinion that a 4-m telescope should be considered. A central question in this regard is whether any sites exist where the seeing is good enough to justify installation of a very large reflector. We understood that in Yünnan Province sites are being tested by Kunming astronomers, who showed a considerable interest in the subject of astronomical site selection. A summary of presentations on this subject made by our delegation to the Kunming astronomers is presented in Appendix D. We gathered that Yünnan is intended to be the principal observatory in the south of China, much as Peking Observatory will be in the north.

Radio Instrument Development

While expansion of optical-observing facilities appears likely, centered on Peking and Kunming, new radio astronomy developments for the present and near future appear to be concentrated at Mi-yun Station of Peking Observatory, where an expansion of the east-west array of reflectors is under way. In addition to the 16 9-m parabolic dishes now in operation, 12 additional 9-m reflectors are under construction. Each of these will be connected to each of the 16 existing antennas to yield an aperture synthesis array with 192 east-west baselines, at 6-m intervals from 12-m to 1,200-m. The antennas will have electrical drive in both hour angle and declination. The array will operate at

176 MHz, using the old 146-MHz transmission line system for local oscillator distribution. There will be a 176-MHz paramp at each antenna. Since only a 96 correlator back-end will be available, it will be necessary to observe a given region twice, each time with 6 × 16 baselines, to do a full synthesis. Correlator outputs will go to an online minicomputer for processing. The specific computer has not yet been selected. A new, modern control building has been completed to house the array controls, minicomputer, electronics, etc. The array should be completed in about 2 years. It will have a 10° field of view and give resolution of 3.8 arc min and sensitivity of 0.15 Jy in a 12-hour observation. This instrument, when it is completed, will probably be the most sophisticated and competitive (with Western instruments) astronomical instrument in China for some time. Its sensitivity is comparable to that being obtained anywhere, at these frequencies.

A second aperture synthesis array is also under construction at Mi-yun. It will consist of 20 dishes, 1.5 m in diameter, operating at 3-cm λ. The array will be on an east-west line, about 100 m in length.

PHYSICAL PLANT EXPANSION

Besides the acquisition or fabrication of the new reflectors discussed so far, expansion of physical plant facilities have been initiated. As is the case for the telescopes, these are indicative of the newly implemented policies to strengthen science in China.

At Hsing-lung Station of Peking Observatory, our delegation visited a newly constructed guest-house where visiting astronomers will be lodged.

A major physical plant enlargement has been started at Yünnan Observatory. On a low-rolling elevation about half a dozen miles outside Kunming an extensive building program has been initiated over an area of 230,000 m². The buildings will accommodate five observational subdivisions of the observatory:

- Solar physics
- Satellite tracking and celestial mechanics
- Astrophysics, including the 1-m telescope
- Astrometry
- Meridian transit and time service

Radio-astronomical techniques will be included in the solar physics and astrophysics and subdivisions. Support facilities will include a computer for analytical work, a library, laboratories, and lecture halls. We estimate that the new buildings will provide space for some 500 staff members.

Although in the immediate future this extensive facility will house all of the instruments operated by Yünnan Observatory, we understood that in the future outlying observing stations at the most favorable sites to be found will be served from this central facility at Kunming.

7

EDUCATION AND RESEARCH IN
ASTRONOMY AT UNIVERSITIES

GENERAL EDUCATION

As the history of general education in China during the past two
decades has greatly influenced the level of education of Chinese astronomers, we review that history here, though in highly summarized
form.

Prior to 1965 the successive levels of education consisted generally
of the following sequence: 6 years of grammar school; 6 years of
middle school (including approximately our high school level); 4 years
of university (corresponding to our undergraduate education) with 1,
or even 2, additional years for some specialities; and 3, or even 4,
years of postgraduate studies consisting mainly of supervised studies
and research (roughly corresponding to our graduate schools).

During the 1950's the Chinese university system was substantially
affected by a government policy aimed at increased effectiveness and
economy by reducing the number of universities that offered degree
courses in fields involving small numbers of teachers and students.
A relevant example of this policy is the formation of the astronomy
department at Nanking University in 1952 by combining there the former
astronomy departments of Chung-shan University in Canton and of Chi-lu
University in Shantung Province with the astronomical subsection previously existing in the physics department in Nanking University.

One of the aims of the Cultural Revolution, starting in 1966, was
to integrate the highly educated fraction of the people into the
general national life by requiring of them 2 or more years of work in
field or factory. To free the necessary time, grammar schools and
middle schools were shortened in most places to last a total of only
10, or at most, 11 years, followed by a minimum of several years assignment to farm or factory work. Simultaneously the majority of university
courses were shortened to 3 years. Eventually the general policy turned
actively against education not directly linked to industrial or agricultural production, so that by 1976 not only were the just-mentioned
reductions in educational years maintained, but essentially all post-
graduate studies had been abandoned. The effectiveness of university
education was further diminished by dropping academic examinations as
prerequisites for entrance into a university and basing the selection
of university students largely on the recommendations of members of the
agricultural communes or factories in which the candidates were working.

Since the decisive turn of political events late in 1976, an extensive discussion has been under way regarding the strengthening of higher education. Concrete proof of the serious intent of this discussion is the fact that both Peking and Nanking universities did not admit new students at the normal time in the fall of 1977, but postponed these admissions to February 1978 for the purpose of gaining the necessary time to make decisions on new admissions policies. By the time of our visit, it appeared assured that academic examinations would again be a major factor in university admissions. Furthermore, during the weeks of our visit it became increasingly clear that at least the major universities would lengthen their regular courses from 3 to 4 years and in some subjects even to 5 years.

Regarding graduate studies, we heard during our visit at Peking University that plans were already being discussed there to include 1,500 students engaged in 3-year graduate programs to be developed in a stepwise buildup over about 2 years. Consistent with general policies, the completion of any of these graduate study programs will not be given any formal designation equivalent to "Ph.D." Just at the end of our visit in China, it was officially announced that the Academy of Sciences had directed the Chinese University of Science and Technology (the only university directly under the academy) to start now 3-year "post-graduate" courses in a number of subjects, and to build up these courses in the near future to a level of about 1,000 students. Only 3 weeks later official announcements were published in which other universities were also authorized and directed by the government to develop graduate study programs in the immediate future.

PROFESSIONAL EDUCATION IN ASTRONOMY

The largest schools of astronomy in China are located in Nanking and Peking. Nanking University contains an independent astronomy department with an instructional staff of about 50. The department has some 120 students, of which 30 to 40 graduate each year. Instruction is largely concentrated on solar and stellar astrophysics, radio astronomy, astrometry, and celestial mechanics. For the first $1\frac{1}{2}$ years the students of the department pursue a common course, which includes a substantial fraction of mathematics and physics courses, and thereafter specialize in a particular branch of astronomy in which they write a thesis during their last year. A substantial fraction of the staff members in the department are active in research. Relations between the department and nearby Purple Mountain Observatory seem to be very good, with highly positive effects on the training of advanced students and on the research of the Department's staff.

Peking Normal University contains an astronomy department. We did not visit it, but we understand that it is not research-oriented in its training course or in its staff's main activities. In contrast, Peking University contains an active school of astronomy as a "specialty" in the geophysics department. This specialty is manned by 17 faculty members of which about one-half are active researchers. The number of students now enrolled in this specialty is 52. However, it was expected that 20 new students were to be admitted in February, indicating a rise in the total astronomy student body of this school in the next 4 years and an eventual annual output of at least 15 astronomy graduates. The students enrolled in the astronomy specialty are required to take courses in mathematics, physics, electrical engineering, and English. Courses concentrate on solar physics, radio astronomy, and stellar physics. Relations between the astronomy section of Peking University and the various branches of Peking Observatory do not seem as well developed as might be optimum for the training of students and for the research of the section's staff.

Some future astronomers are being trained also at other universities, and research in astronomy is being carried out in various departments in a number of other universities and institutions. Specifically, at the Chinese University of Science and Technology several members of the physics department are actively engaged in theoretical research in modern astrophysics. In most cases relations between these researchers and the observational research at active observatories appear to be tenuous. In this connection the future, potentially important, role of Yünnan Observatory appears presently unclear.

Our question as to whether the total annual output of new astronomers--the majority of which are not likely to have an opportunity for formal graduate studies--could be absorbed by astronomical institutions in the future was answered affirmatively.

Altogether, we feel that the effectiveness of university training in astronomy in China might be substantially enhanced if the teaching staff could actively participate in stimulating modern research, observational or theoretical, to a larger degree than seems to be normally the case at present.

EDUCATIONAL BACKGROUND OF PRESENT RESEARCH ASTRONOMERS

The majority of the active research astronomers whom we met appeared to be between 35 and 50 years old. This age-group had by and large finished its formal education before 1965 and therefore had the advantage of having gone through the earlier full Chinese education schedule, which, measured by the total number of years, is quite comparable to that of many countries in which substantial astronomical research is carried out. At the same time, this age-group has not had the opportunity of broad and reasonably frequent contacts with leading research astronomers from other countries--notwithstanding outstandingly effective exceptions like Dr. Christiansen and Dr. C. C. Lin. We found that Chinese astronomers make very thorough use of the astronomical world literature, which is quite generally and promptly available to them. However, our detailed research discussions clearly indicated, as might be expected, that reading scientific papers is not a full substitute for direct meetings in which new ideas and new developments of thought can be exchanged years before printed publication.

The absence of the age-group between 25 and 35 among active Chinese research astronomers is striking and was often emphasized to us by leading Chinese astronomers. How serious for Chinese astronomy the lack of this age-group will turn out to be will greatly depend on the speed with which higher education in science and technology in general, and in astronomy in particular, will now develop. On this point the prospects brightened during our visit in a breathtaking manner.

8

DISCUSSION AND ISSUES

INTERACTION BETWEEN CHINESE ASTRONOMERS AND THE INTERNATIONAL
COMMUNITY

The People's Republic of China and the International Astronomical Union

The continued absence of the People's Republic of China from partici-
pation in the International Astronomical Union (IAU) is a severe loss
to the world community of astronomers. The problem at issue, which is
still a perplexing one, is the membership of Taiwan, which the PRC
considers unacceptable. By way of background, China joined the IAU in
1935 and was represented at the general assemblies of 1938 by Tai
Wen-sai and of 1948 by Ch'eng Mao-lan, the latter then resident in
France. After the founding of the People's Republic, China ceased to
participate in the work of the IAU until its then President Otto Struve
invited the astronomers from the PRC to resume the representation of
China at the Dublin General Assembly in 1955. Four PRC astronomers
were in attendance at Dublin and seven at Moscow in 1958.

In 1959, when Taiwan applied and was accepted for membership, the
Executive Committee of the IAU specified that it would represent only
the geographical territory of Taiwan, but the PRC nevertheless withdrew
from the union early in 1960 and has firmly refused to reconsider un-
less Taiwan is expelled. It is the position of the PRC that, since
Taiwan is a province of China, the Chinese Academy of Sciences (CAS)
of Peking represents astronomers on Taiwan as well as on the mainland.
The IAU holds that the CAS cannot represent Taiwan astronomers as long
as a government other than that of the PRC controls the territory of
the province. Therefore, it feels that to deny membership to Taiwan
would exclude its scientists from taking part in the activities of the
IAU, thereby violating one of the union's most basic principles, that
of the universality of science.

The Chinese astronomers and administrators with whom we discussed
the IAU, while unyielding in their attitudes on the Taiwan question,
seemed to agree that the impasse was a matter of differing philosophical
principles and was not in any way motivated by lack of goodwill on the
part of either their country or the IAU. We believe that until the impasse is resolved, we should do everything possible to promote the

development of cooperative projects between China and other countries and to encourage attendance by the Chinese at international conferences that are not sponsored by the IAU.

Chinese Astronomical Society

The Chinese Astronomical Society has been inactive for about 15 years. In the past, it held astronomical meetings, both central and regional, at which papers were presented. It has been difficult to find out how its operations, when it was active, compared with those of other national societies, for example, the American Astronomical Society. Burbidge, as the current president of the American Astronomical Society, presented messages of friendship and encouragement to our sister society, though there were only two people she could locate who had some official standing in the Chinese Astronomical Society: its president, Chang Yü-che, Director of Purple Mountain Observatory, and Wang Shou-kuan of Peking Observatory, who is a councilor.

Wang said he had been a councilor for a very long time, because there had been no elections since the society became inactive about 15 years ago. According to what we were told, it used to hold both central and more frequent regional meetings. It does not appear that it ever had responsibility for publishing astronomical journals, as the American Astronomical Society does.

Wang said he believed the society would be regenerated shortly, and information has come to a member of the delegation that a "scholarly annual meeting" of the Society was held recently in Shanghai.

International Exchange Between Astronomers in the United States and China

The visit by the Chinese delegation of astronomers in 1976, and our visit to China have opened the way to a continuing exchange of information and ideas with China and, it is to be hoped, of astronomers.

The science of astronomy has always depended on worldwide cooperation, a fact well known and widely recognized when the International Astronomical Union was set up after World War I. Many fundamental astronomical programs are long term and require data from different instruments spaced around the world. In the past, astrometric, solar, and solar-system programs were prime examples of this kind of cooperative work. More recently, the very-long-baseline interferometric work at radio wavelengths is a prime example, where the present long baseline is essentially the largest fraction of the Earth's circumference over which cooperative programs can be organized. In the future, many programs in space astronomy may involve worldwide collaboration.

Many older-generation Chinese astronomers have spent time in the United States and Europe. Some of the middle-generation astronomers were abroad at the time of Liberation and have since returned to China. But the younger generation have had no experience of astronomy in

Europe and the United States. The recognized world leadership in observational astronomy--especially in large telescopes with modern instruments in good climates--is in North and South America. The Chinese are on the point of entering this worldwide community, with the construction of their 2-m telescope, their consideration of a larger telescope, and their readiness to look into the possibility of further astronomical ventures in the future.

China's millenia-long tradition in the science of astronomy, her strong groups of theoretical workers, and her present recognition that the solutions to the fundamental problems of the nature of matter, energy, the laws of physics, and the universe may be sought in astronomy will be powerful pointers to the need for international collaboration.

We believe that such international collaboration should be sought and encouraged. There are some obvious ways of doing this. Invitations should be issued to established Chinese astronomers to spend time at major observational research institutions in the United States, where they could compare and evaluate different types of modern auxiliary instrumentation on telescopes and could take part in discussions on future ground-based and space telescopes and astronomical programs.

It could be made known that American astronomers would welcome invitations to spend time at Chinese observatories and astronomical institutions where new instrumental developments are under way. Above all, we welcome plans now underway to invite Chinese students and young workers (post- or predoctoral) to major American astronomical institutions. However free the exchange of literature, correspondence, and information, there is no substitute for personal contact. We learned this during our visit to China, but 4 weeks is all too short a time and longer visits--in both directions--are needed.

In the longer term, if China's present drive toward basic science and technology is maintained, it might be possible, when the 2-m telescope and perhaps a larger aperture instrument come into operation, to arrange for exchange of observing time. No such arrangement exists at present between the United States and other countries, but two international ventures are about to come into operation--the French-Canadian 3.6-m telescope on Mauna Kea, Hawaii, and the British infrared telescope. Both of these involve the provision of observing time to the University of Hawaii in return for the sites for these telescopes.

OPTIONS FOR FUTURE DEVELOPMENT OF OPTICAL INSTRUMENTS

The Chinese engineers and astronomers were reluctant to discuss future ventures such as the construction of a 4-m reflector, a possibility raised by the delegation more than once. Perhaps the reason for their hesitation was that such planning is the responsibility of others, or possibly because the 2-m telescope was already such a major step for them to take that they felt it best to gain experience with it before considering future options. We feel that a central early requirement for any planning for a large telescope in China should consist of a major site-testing program. We understand that such a program is presently under way in Yünnan Province, but did not hear of parallel undertakings in other parts of China. The question of the location and quality of the best astronomical sites available in a large continental landmass such as China is not an easy one according to recent Western experience.[1,2] The answer to this question might influence the size, character, and width of purpose--general or specialized--of the next larger Chinese telescope.

OPTIONS FOR FUTURE DEVELOPMENT OF RADIO ASTRONOMY INSTRUMENTS

The most obvious possibility for further development of radio astronomy instruments is expansion of the Mi-yun 176 MHz array. This instrument could easily be expanded by adding a north-south arm to give better beam characteristics at low declinations and/or lengthening the east-west arm to give higher resolution. Additional antennas would also increase the sensitivity somewhat, as would improved amplifiers. There are at present relatively few radio telescopes operating at frequencies less than 300 MHz. It is a region of the spectrum where observations with greater resolution and sensitivity than presently available are badly needed. The array now under construction should make a significant contribution; an expanded version could be the best instrument in the world in this frequency range.

Chinese radio astronomers appear well aware of these opportunities. However, since their present building program is still 2 years from completion, they were understandably reluctant to discuss the next stage of development in any detail.

Another, possibly feasible, area for further development is millimeter wavelength instrumentation. Some development work at 8-m wavelengths is already being done at Purple Mountain Observatory (see Chapter 5). In addition, astronomers at Peking University have designed an interesting radio heliograph for millimeter wavelengths. It consists of an equatorially mounted beam, 9.2-m long, on which are placed eight parabolic reflectors, each 40 cm in diameter. The beam is rotated at one revolution per 10 minutes in a plane perpendicular to the line of sight. This yields a synthesized map of the Sun every 5 minutes, with a resolution of about 1 arc min. at 4-mm wavelength. It appears to be a good conceptual design, and it would be a useful instrument for solar work if it were built. At present there is no instrument in the world that regularly observes the Sun at millimeter wavelengths with this high resolution.

In general, further development of radio astronomy in China is critically dependent on the availability of sensitive amplifiers and detectors. We were told that there are no parametric or cooled amplifiers in China. Since these are of some importance in various applied fields, as well as in radio astronomy, their development in the near future appears logical.

THEORETICAL ASTROPHYSICS

Among the Chinese astronomers we met or learned about through presentations and discussions were a large number of theoreticians. Many of these are located at the observatories and universities of Peking and Nanking, but quite a number of them are working in other institutions, such as the physics department of the Chinese University of Science and Technology. We were impressed by the very substantial number of theoreticians who seemed to be of truly high caliber. Taken as a group their research covered a wide range of central problem areas of modern astronomy.

Though we found the state of theoretical astronomy in China to be generally promising, we would like to point out several issues that we feel may affect the long-term development of Chinese astronomy. First, the electronic computers available to the theoreticians are moderate in speed and capacity and may eventually affect productivity, though this is not now a major handicap. A more serious problem is the scarcity of opportunity for direct discussions with researchers in other countries working on the same or related subjects (see Chapter 7).

We are primarily concerned, however, about the minimal level of active research contact between theoretical and observational astronomers in China, though there are some notable exceptions. This is understandable--the bright theoreticians are naturally drawn to topics that are at the extreme forefront of modern astrophysics, and it is these topics that, for relevant new observations, require larger telescopes than are presently available in China. Obviously there are exceptions to this general statement. In fact, during our visit there were some lively discussions about a few observational programs that could be executed with existing Chinese telescopes and that would be of direct interest in certain important, though limited, problems in modern theoretical astrophysics.

We feel that a major step towards increased interaction between theorists and observers could be achieved very quickly if some of the bright theoreticians would direct at least part of their research to the most important of those topics for which observational data can be obtained with existing telescopes. At the same time, it seems clear that a full range of interaction between theory and observations will open up only when newer and larger instruments become available. It is for this reason that the 2-m telescope now being constructed for Peking Observatory assumes in our judgment such an important role. Though one should view it presumably as only a first step, it is large enough to enable imaginative observers to derive new observational data relevant to exciting topics in modern astronomy that would attract bright theoreticians. It thus could provide an effective basis for a more general interaction between observational and theoretical astronomers. Accordingly, we feel that the earliest possible completion of this telescope would be of extraordinary importance.

In addition, there is one other step that would enhance stronger interaction between researchers, at least in the Peking area, in all branches of astronomy. This would be provision of a new headquarters for Peking Observatory, large enough to provide adequate office space for a research center, not only for the staff members of all the branches of the observatory, but also for the research astronomers of Peking University and for guests from other neighboring institutions. We believe that a center for common discussion and interaction could greatly enhance the effectiveness of astronomical research in the Peking area through closer and more frequent interaction between solar, stellar, radio, and theoretical astronomers.

PUBLICATION PROBLEMS

At present there appear to be only two regular publication outlets for Chinese astronomy: *Acta Astronomica Sinica* and *Scientia Sinica*. Both are available in at least some libraries outside the country.

Acta began publishing from Purple Mountain Observatory in the early 1950's, appearing with annual volumes each containing two issues of about a dozen articles. Interrupted during the Cultural Revolution, it resumed publication with Vol. 15 in 1974 and now combines for China the roles of our *Astronomical* and *Astrophysical Journals*.

Scientia is a publication of the CAS. It resembles our *Proceedings of the National Academy of Sciences*, and like ours carries occasional articles on astronomy.

Individual astronomy centers produce articles that in some cases may be printed but in most cases are merely typewritten and duplicated. These appear to circulate in China partly as preprints, partly in lieu of formal publication, but do not seem to be distributed significantly outside the country. In the West, the extensive circulation of preprints on an international scale has had important benefits in generating constructive early criticism and in speeding the dissemination of the latest research results to active investigators. Membership in the network of preprint distribution could help materially in bringing Chinese astronomical research into the mainstream of world astronomy.

Finally, a language barrier remains. Essentially all Chinese astronomers appear to have some reading knowledge of English, but the reverse is scarcely the case. In view of the growing amount of important astronomical research being published in China, coupled with the difficulties and at best the delays in getting full translations into Western languages, Chinese authors and editors should be encouraged to pay particular attention to the translated abstracts that normally appear in *Acta* and any other Chinese source. These abstracts should be relatively detailed and quite specific in giving conclusions and numerical results. A common problem with abstracts in any language is the tendency simply to repeat the title in a wordier way, or to give only vague generalities about the work. When the article is in a familiar language the critical information can still be retrieved. But with a language as little understood by most astronomers as Chinese, an inadequate abstract causes the article simply to be ignored.

Conversely, with a good abstract the salient points can be referenced, and if the matter is important a full translation will be obtained or at least eagerly awaited in the translation journal. Concentration of essentially all astronomy in the *Acta*, along with its full translation, as recently begun by T. Kiang, would thus ensure worldwide availability and use of Chinese astronomy.

REFERENCES

1. B. Innes, M. Hartley, and T. T. Gough, "Astronomical Seeing and Meteorological Air Mass Analysis," *Obs.*, 1974, 94:14.
2. M. F. Walker, "Report of the I.A.U. Commission 50, Sec.3: Identification of Potential Observing Sites," *Trans. I.A.U.*, 1976, 16A, pt. 1, 222.

APPENDIX A: PRINCIPAL INSTRUMENTS, RESEARCH PROGRAMS, AND STAFF OF THE VARIOUS INSTITUTES AND UNIVERSITY DEPARTMENTS

PEKING OBSERVATORY--SHA-HO STATION

Instruments

1. 60-cm solar coudé reflector (made in Nanking).
2. 13-cm optical white-light refractor for sunspots.
3. 6-cm chromospheric refractor (made in Russia; 25-cm aperture with Hα filter, installed in 1957).
4. 14-cm photoelectric transit instrument (for time).
5. Photoelectric astrolabe (variation of Earth's rotation, latitude; installed in 1963).
6. Two 2-m radio dishes working at 3 cm and 10 cm for solar patrol.

Principal Research

1. Solar, for forecasting events. Involved are flare patrol, chromospheric patrol, radio patrol.
2. Solar physics. Structure of flares as related to magnetic fields of sunspots. Radio solar physics to study structure of the events.
3. Time service. Involved are transit instrument, photoelectric astrolabe.
4. Astrometric. Rotation of Earth, variation of latitude (polar wander).

Observing Conditions

Two hundred and fifty days clear enough for solar work; 200 nights for transit and astrolabe.

PEKING OBSERVATORY--HSING-LUNG STATION

Instruments

1. 60-cm reflector (made in Nanking, installed in 1968).

2. 60/90-cm Schmidt (Zeiss Jena ~1968?).
3. 40-cm double astrograph (Zeiss Jena 1974).

Principal Research

1. Photoelectric study of binaries, eruptive variables (reflector).
2. Flare star search (Schmidt, astrograph).
3. Nova spectra (Schmidt in Nasmyth mode).
4. Stellar evolution.
5. Theory of structure and evolution of galaxies.

Observing Conditions

Two hundred clear nights per year; rainfall ~600 mm/yr. Seeing can be as good as ~1". Bad seeing in winter. Typical wind speed ~3 m/s. Must close often in spring due to dust.

PEKING OBSERVATORY--MI-YUN RADIO ASTRONOMY STATION

Instruments

Solar radio astronomy to be extended to galactic when interferometer is expanded again in 1978.

PEKING OBSERVATORY--TIENTSIN LATITUDE STATION

This is the principal latitude station for China.

Instruments

1. Photoelectric astrolabe.
2. Zenith tube (made in Nanking; described in *Acta Astronomica Sinica*, vol. 18, p. 38).

Research

Polar wander. Variations of Earth's rotation rate.

PRINCIPAL PERSONNEL--PEKING OBSERVATORY 北京天文台

Peking Administrative Offices

Ch'eng Mao-lan   Overall Scientific Director

Yü Chiang		Chief Administrator
Hung Ssu-yi	洪斯溢	Secretary and Administrator

Sha-ho Station

Liang Hsueh-tseng		Director
Lo Ting-chiang		Astrometry
Shih Chung-hsien	史忠先	Solar Physics
Ch'ien Shan-chieh		Radio Astronomy
Chang Pao-ts'ai		Astrometry
Po Shu-jen		History of Science
Li Ch'i-pin	李启斌	Theoretician

Hsing-lung Station

Cheng Yüan-chang	郑元章	Director
Huang Lin	黄 磷	Stellar Physics, Chief Scientist
Shen Liang-tsao		Stellar Physics
Ch'en Chien-sheng	陈建生	Stellar Physics
Li Feng	李 峰	Stellar Physics
Chai Ti-sheng	翟迪生	Stellar Astrophysics
Yin Chi-sheng	尹济生	Stellar Astrophysics
Tu Po-t'ien	杜柏田	Stellar Astrophysics
Hsu Te-hui	徐德辉	Stellar Astrophysics
Yeh Chi-t'ang	叶基棠	Stellar Astrophysics
Chang Chi-t'ung	张济同	Stellar Astrophysics
Chiang Shih-yang	蒋世仰	Stellar Astrophysics
Sun Yi-li	孙益礼	Stellar Astrophysics

Kuo Hsiao-chen	郭筱贞	Stellar Division
Hsia Ch'en-tao	夏臣道	Stellar Division
Mei Pao	梅 苞	Stellar Division
Liu Tsung-li	刘宗礼	Stellar Division
Shen Liang-yi	沈良义	
Huang Hung-pin	黄鸿宾	Administrator
Mi-yun Station	密云天文台	
Wang Shou-kuan	王绶琯	Radioastronomy, Director
Lo Shao-kuang	罗绍光	Instructor
Lo Yü-sheng	罗雨生	Radio
Ch'iu Yü-hai	邱育海	Astronomy
Liu Hsü-chao	刘绪昭	Astronomy
P'iao T'ing-yi	朴廷彝	Radio and Communication
Chang Hsi-chen	张喜镇	Astronomy
Yen Feng-kao	阎凤高	Administration
K'ang Lien-sheng	康连生	Astronomy

YÜNNAN OBSERVATORY, KUNMING

Instruments

1. 13-cm refractor (sunspot mapping).
2. 14-and-one-half-cm chromospheric refractor (Hα solar photographs).
3. 40-cm solar spectroheliograph (simultaneous photographs in 10 wavelengths).
4. Satellite tracking instrument (did not see, but described as a first-generation visual instrument plus a geodetic theodolite).
5. New satellite tracker to be obtained, like the one we saw in Nanking Instruments Factory.
6. New 10-m Zeiss Jena reflector to be installed 1978.
7. Photoelectric transit.

Principal Research

1. Solar physics: flare prediction by mapping sunspots, chromospheric and flare patrol, magnetic field measurements by Leighton photographic subtraction method, spectroheliograms.
2. Celestial mechanics related to satellites and their tracking.
3. Astrometry. Measurement of time; measuring polar motion. Part of all China net in these two subjects.
4. Stellar physics will begin with installation of 1-m reflector in 1978.
5. History via ancient Chinese records.

Observing Conditions

Average temperature 15° C. Altitude 2,020 m. Close to the lights of Kunming. Have 215 completely clear days per year. Transparency excellent (coronal blue). Can make solar observations of some kind on 310 days per year. Solar seeing often excellent (< 1") for very short periods. They need a better site for very large nighttime telescope, away from Kunming lights.

INDIVIDUALS MEETING GROUP AT KUNMING AIRPORT 昆明飞机场

Yü Fu-t'ing	于鲮亭	Deputy Director, Science and Technology Office, Provincial Revolutionary Committee
Wu Min-jan	吴敏然	Director, Yünnan Observatory
Lin Chao-chü	林兆驹	Astronomer, Yünnan Observatory
Ting Yu-chi	丁有济	Astronomer, Yünnan Observatory
Li Chih (F)	栗 志	Technical Group, Yünnan Observatory

PRINCIPAL PERSONNEL - YÜNNAN OBSERVATORY 昆明天文台

Yü Fu-t'ing	于鲮亭	Deputy Director of Yünnan Provincial Committee of Science and Technology
Wu Min-jan	吴敏然	Director
Lin Chao-chü	林兆驹	Celestial Mechanics, Dynamics
Chang Pai-jung	张柏荣	Solar Physics (will be in charge of new 1-m)

Ting Yu-chi	丁有济	Solar Physics
Huang Yin-liang	黄寅亮	Density wave theory. Preparatory group for 1-m. Site testing for larger telescope.
Huang Jun-ch'ien	黄润乾	Preparatory Group
Chiang Ch'ung-kuo	姜崇国	Astro Survey
Nieh Chao-ming	聂昭明	Stellar Physics
Chou Yün-fen	周允芬	Solar Physics
Li Chih	栗 志	Technical Group

INDIVIDUALS WHO ATTENDED OUR LECTURES IN KUNMING, OCTOBER 11, 1977

Yünnan Observatory	云南天文台	
Huang Yin-liang	黄寅亮	Stellar Physics Preparatory Group
Chang Chu-wen	张筑文	Solar Physics
Lo Pao-jung	罗葆荣	Solar Physics
Wu Ming-ch'an	吴铭蟾	Solar Physics
Lin Chao-chü	林兆驹	Administration
Chang Chu-heng	张竹恒	Radio
Pao Meng-hsien	鲍梦贤	Stellar Physics Preparatory Group
Chiang Ch'ung-kuo	姜崇国	Astro Survey
Nieh Chao-ming	聂昭明	Celestial Mechanics
Liu Chih-huang	刘之煌	Solar Physics
Ku Hsiao-ma	顾啸马	Solar Physics
Li Chih (F)	栗 志	Technical Group
Chang Heng	张 衡	Solar Physics
Huang Jun-ch'ien	黄润乾	Stellar Physics Preparatory Group
Yin Shu-hua	尹淑华	Solar Physics

Chiang Chih-chung	蒋治中	Solar Physics
Chou Yün-fen (F)	周允芬	Solar Physics
Yang Jung-pang	杨荣邦	Solar Radio
Yang Cheng-k'uei	杨正逵	Research Affairs Department
Ch'in Sung-nien	秦松年	Stellar Physics
Hsieh Kuang-chung	谢光中	Stellar Physics Preparatory Group
Yünnan University	云南大学	
Tseng Chao-ch'üan	曾昭权	Physics Department
Wang Chung-yung	王仲永	Physics Department
Tseng Chao-yi	曾昭义	Physics Department
Hsiung Yeh	熊 烨	Physics Department
Chao Shu-sung	赵树松	Physics Department
Wang Chih-kuei	王志珪	Geophysics Department
P'eng Shou-li	彭守礼	Physics Department
Chang Shih-chieh	张世杰	Physics Department
Li Ho-nien	李鹤年	Physics Department
Ch'en Chung-hsüan	陈中轩	Physics Department
Yü Ch'uan-tsan	喻传赞	Physics Department
Cheng T'ien-min	郑天民	Physics Department
Chao Chao-wang	赵昭旺	Stellar Physics
Ch'ü Chien-hsün	瞿建勋	Technical Research
Su Pu-mei	苏步美	Technical Research
Ling Tsung-hsü	凌宗顼	Materials
Teng Li-wu	邓立吾	Stellar Physics
Liao Yüan-chin	廖远金	Solar Radio

| Chou Jen-ming | 周仁铭 | Physics Department |
| Mu Chün | 木 钧 | Physics Department |

Kunming Teacher's College　昆明师范学院

| Yang Ch'ing-wen | 杨庆文 | Physics Department |
| Ch'en T'ing-chin | 陈庭金 | Physics Department |

SHANGHAI OBSERVATORY

Instruments

　　1.　Photoelectric astrolabe (installed 1974; latitude, time).
　　2.　Transit (for time only).
　　3.　Visual astrolabe (French, installed 1958).
　　4.　Clocks (quartz, atomic, shock wave).
　　5.　Experimental radio interferometer.
　　6.　40-cm double astrograph (1900 from Jesuit college) at Zo-se.
Scale ~35"/mm.

Principal Research

　　1.　Time service (polar motion, continental drift, Earth's rotation,
time comparisons).　Shanghai is center of the Chinese time service.
　　2.　Comparison of time standards.
　　3.　Transmission of Chinese radio time signals.
　　4.　Photographic astrometry, using first-epoch plates of Orion
Nebula, M67, M8, NGC 2530; recently repeated for third epoch.　Second-
epoch proper motions published ~1960 by Li Hen of the staff.
　　5.　Theory:　high-energy astrophysics, galaxy distributions.
　　6.　Experimental tests of radio interferometry for improved time
determinations.

Observing Conditions

The astrometric instruments in Shanghai can be used between 100 and 150
night per year.

PRINCIPAL PERSONNEL - SHANGHAI OBSERVATORY　上海天文台

| Chou Tsun-po | 周尊博 | Director |

Wu Lin-ta	邬林达	Radio Astronomy
Chuang Ch'i-hsiang	庄奇祥	Time Frequency
Tu Hsiu-feng	杜秀峰	Time Frequency
Chao Chün-liang	赵君亮	Celestral Physics
Yen Lin-shan	阎林山	Celestral Survey
Hu Chung-wei	胡中为	Evolution of Solar System
Wan T'ung-shan	万同山	Celestral Survey
Chin Wen-chiao	金文教	Universal Time
Hsueh Tao-yüan	薛道远	History of Astronomy
Ts'ai Yung-ming	蔡永明	Library
Yü Kuo-chen	于国桢	Administrator
Tai Wen-sai	戴文赛	Evolution of Galaxy and Solar System, University of Nanking

PURPLE MOUNTAIN OBSERVATORY

Instruments

Solar Physics

1. Two-m radio telescope for 3-cm and 10-cm work (China).
2. 40-cm horizontal spectroheliograph for 11 simultaneous mono-chromatic pictures (China).
3. 16-cm chromospheric refractor (Secasi, Bordeaux, France, 1957).

Stellar Physics

1. 60-cm reflector (Zeiss Jena ~1934, restored 1957).
2. 40-cm double astrograph (Zeiss Jena 1964).
3. 43/60-cm Schmidt (first Chinese-made large telescope, in ~1964).

Time Service

1. Transit (started work in 1957).
2. Photoelectric transit (1964).

Satellite Trackers

1. 15-cm tracking theodolite.
2. Also used the 43/60-cm Schmidt.
3. Will have Doppler velocity tracker.

Principal Research and Professional Activities:

1. Time service, started in 1957 (first in modern China). Teamed with Shanghai Observatory in 1959 to start the National Time Service.
2. Satellite tracking. Obtained positions and calculated orbits of first two Chinese satellites (of 1970 and 1971).
3. Solar physics: predictions of flares; theory of solar activity.
4. Stellar physics: short-period variable stars in globular clusters; flare stars; spectrophotometric gradients.
5. Almanac. Began in 1964 to compile, compute, and publish the *Chinese National Almanac and Ephemeris*.
6. Planetary. Together with Hsing-lung Peking Station, independently discovered ring of Uranus.
7. Comets. Discovered two in ~1973 and computed their orbits.
8. Theory: high-energy astrophysics, stellar evolution of late stages, density wave theory.
9. History of astronomy in ancient China.

Observing Conditions

Strong lights of nearby Nanking (about 1 magnitude increase in sky surface brightness). Seeing generally 2" or larger.

PRINCIPAL PERSONNEL--PURPLE MOUNTAIN OBSERVATORY　紫金山天文台

Chang Yü-che	张钰哲	Director; Solar System Astronomy, History
Chao Wen-piao	赵文彪	Administrator
Hsiang Te-lin	向德琳	Administrator; Radio Astronomer
Kung Shu-mo	龚树模	Head, Stellar
Kuo Ch'üan-shih	郭权世	Head, Solar
Wang Ching-sheng	王京生	Head, Radio
Wang Jung-ch'uan	王荣川	Head, Satellites
Chang P'ei-yü	张培瑜	Head, History

Cheng Ying	郑 莹	Head, Practical Astronomy
Yü Tsung-k'uan	余宗宽	Head, Ephemeris
Yang Shih-chieh	杨世杰	Head, Instrumentation

INDIVIDUALS PRESENT DURING BRIEFING OR LECTURES AT PURPLE MOUNTAIN OBSERVATORY, OCTOBER 16, 1977

Purple Mountain Observatory	紫金山天文台	
Kuo Ch'üan-shih	郭权世	Head, Solar Physics
Yang Shih-chieh	杨世杰	Head, Instrumentation
Wang Ching-sheng	王京生	Head, Radio Astronomy
Chang P'ei-yü	张培瑜	Head, Ancient History of Astronomy
Cheng Ying	郑 莹	Head, Applied Astronomy
Yü Tsung-k'uan	余宗宽	Head, Ephemeris
Wang Jung-ch'uan	王荣川	Head, Satellite
Hsiang Te-lin	向德琳	Secretariat, Radio Astronomy
Tseng Wu-chu	曾务珠	Head, Secretariat
Kung Shu-mo	龚树模	Head, Stellar Physics
Chao Wen-piao	赵文彪	Vice-Chairman, Revolutionary Committee
Chang Kuo-hsiang	张国香	
Hsu Ping-ts'ai	许炳才	
Tai Yung	戴 勇	Staff Worker (Cadre)
Sung Chin-yü	宋瑾瑜	Staff Worker (Cadre)
Nanking University	南京大学	
Ts'ui Lien-shu	崔连坚	Solar Physics

Provincial Science and Technology Committee 省科委

Li Yi-min	李益民	Staff Worker
Chang Wen-ming	张文明	Staff Worker

INDIVIDUALS PRESENT DURING VISIT TO NANKING UNIVERSITY,
OCTOBER 17, 1977

Nanking University	南京大学	
Kao Chi-yü	高济宇	Vice-President, Nanking University
Ts'ui Lien-shu	崔连竖	Astronomy Department
Hsü Ao-ao	许敖敖	
Wang Chen-ju	汪珍如	
P'eng Yün-lou	彭云楼	Radio Astronomy
Li Ch'un-sheng	李春生	Radio Astronomy
Huang K'o-liang	黄克谅	Astrophysics
Wang Hsiang-pao	王相宝	Staff Worker, University Revolutionary Committee
Chiang Yao-t'iao	蒋窈窕	Astrophysics
Wang Ch'i-pi	王其碧	Staff Worker
Chou Sung-shan	周松山	Secretary, University Revolutionary Committee
Shih Shih-yüan	施士元	Nuclear Physics
Purple Mountain Observatory	紫金山天文台	
Hsiang Te-lin	向德琳	Secretariat, Radio Astronomy
Kung Shu-mo	龚树模	Stellar Physics

INDIVIDUALS PRESENT DURING VISIT TO PURPLE MOUNTAIN OBSERVATORY,
OCTOBER 18, 1977

Purple Mountain Observatory   紫金山天文台

Huang K'un-yi	黄坤仪	Astrodynamics
Chang Ho-ch'i	张和祺	Stellar Physics
Liu Ts'ai-p'in	刘彩品	Stellar Physics
Ch'u Yü-hua	初毓桦	Stellar Physics
Su Hung-chün	苏洪钧	Stellar Physics
Hang Heng-jung	杭恒荣	Stellar Physics

Chinese University of Science and Technology   中国科技大学

| Chang Chia-lü | 张家铝 | Theoretical Astrophysics Group |
| Yu Chün-han | 尤峻汉 | Theoretical Astrophysics Group |

Nanking Electrical Communication Equipment Factory   南京电讯仪器厂

| Lu T'an | 陆 谈 | Physics |

Shensi Observatory   陕西天文台

| Ch'en Hung-ch'ing | 陈洪卿 | Solar Radio |
| Yao Lien-chih | 姚连芝 | Solar |

PEKING UNIVERSITY

Teaching equipment is principally in radio astronomy using a 2-m solar
radio telescope. We saw no optical instruments for teaching or research.

PROFESSORS AND RESEARCH WORKERS MET AT PEKING UNIVERSITY   北京大学

| Chou P'ei-yüan | 周培源 | President of Peking University |
| Chang Lung-hsiang | 张龙翔 | Professor of Biochemistry and Head of Department of Educational Revolution |

Hsing Chün	邢骏	Director of Astronomy, Department of Geophysics
Yang Hai-shou	杨海寿	Instructor of Astronomy
Wu Hsin-chi	吴新基	Instructor of Astronomy
Lo Tung	罗动	Instructor of Astronomy
Yao Te-yi	姚德一	Instructor of Astronomy
Chao En-p'u	赵恩普	Revolutionary Committee Cadre
P'eng Ch'iu-ho	彭秋和	Faculty?
Yin Ch'i-feng	尹其丰	Faculty?
Ma Erh	马珥	Faculty?
Chou Tao-ch'i	周道祺	Faculty?
Ch'iao Kuo-chün	乔国俊	Faculty?
An Ching-chu	安景竹	Faculty?

NANKING UNIVERSITY

Research subjects now underway by members of the faculty include:

1. Physics of solar activity.
2. High-energy astrophysics.
3. Structure and evolution of galaxies.
4. Polar motion.
5. Celestial mechanics.
6. Orbits of satellites.

Student training equipment includes:

1. 15-cm reflector for *UBV* photoelectric photometry.
2. 28-cm refractor with objective prism, a gift from Harvard College Observatory in 1950.
3. 2-m radio dish for 3-cm patrol of the sun.
4. Zenith telescope for senior astrometry theses.

NANKING UNIVERSITY FACULTY PRESENT AT EVENING SESSION ON THE EDUCATION
OF ASTRONOMERS, AT NANKING HOTEL OCTOBER 8, 1977

Chiang Yao-t'iao　　蒋窈窕

Wang Chen-ju　　汪珍如

P'eng Yün-lou　　彭云楼

Ch'ien Yung-t'ung　　钱永统

Li Ch'un-sheng　　李春生

Huang K'o-liang　　黄克谅

Hsia Yi-fei　　夏一飞

Sun Yi-sui　　孙义燧

Huang Yung-ku'an　　黄永宽

Hsü Ao-ao　　许敖敖

STAFF OF THE NANKING ASTRONOMICAL INSTRUMENTS FACTORY MET DURING EVENING
SESSION AT THE NANKING HOTEL, OCTOBER 19, 1977

Ch'en Hsien-lung　　陈贤龙　　Mechanical Department

Huang T'ieh-ch'in　　黄铁琴　　Mechanical Department

Su Ting-ch'iang　　苏定强　　Optical Department

Chang Che-min　　张哲民　　Administration

Kuo Nai-shu　　郭乃竖　　Mechanical Department

Hao Ch'ing-hsiang　　郝庆祥　　Mechanical Department

Li T'ing　　李挺　　Mechanical Department

Yang Huan　　杨环　　Mechanical Department

Shen ? -an　　沈?安　　Electrical Department

Li Teh-p'ei　　李德培　　Optical Department

Ku Chen-lei　　　　Mechanical Department

Hu Ning-sheng　　　　Director of Design

Wang Ya-nan　　　　Optical Department

INDIVIDUALS PRESENT AT MEETINGS WITH NATHAN SIVIN OCTOBER 6-9, PEKING;
HISTORIANS OF SCIENCE

Cheng Wen-kuang	郑文光	Peking Observatory
Ch'en Chiu-chin	陈久金	Research Institute of the History of Natural Sciences
Ch'en Mei-tung	陈美东	Research Institute of the History of Natural Sciences
Hsi Tse-tsung	席泽宗	Research Institute of the History of Natural Sciences
Hsüan Jui-kuang	禤锐光	Observatory
Yi Shih-t'ung	伊世同	Peking Planetarium
Li Chien-ch'eng	李鉴澄	Peking Planetarium
Liu Chin-ch'i	刘金沂	Research Institute
P'an Chi-hsing	潘吉星	Research Institute
Lu Huai-fa	陆怀发	Research Institute
Po Shu-jen	薄树人	Research Institute
Tu Shih-jan	杜石然	Research Institute
Wang Chien-min	王健民	Research Institute
Wang Li-hsing	王立兴	Peking Observatory
Wang Pao-chüan	王宝娟	Peking Observatory
Yen Tun-chieh	严敦杰	Research Institute
Lu Yang	卢央	Observatory
Huang Wei	黄炜	Research Institute

OCTOBER 7, PEKING LIBRARY

Pao Cheng-ku	鲍正鹄	Associate Director

OCTOBER 14, SHANGHAI; HISTORIANS OF SCIENCE

Ch'üan Ho-chün	全和钧	Shanghai Observatory

Ho Miao-fu	何妙福	Shanghai Normal University
Hsueh Tao-yüan	薛道远	Observatory
Hsü T'ien-fen [?]	徐天芬	University
Kung Hui-jen	龚惠人	Observatory
P'an Nai	潘鼐	Preparatory Group for Shanghai Planetarium
Wang Hsien-chu	王先铸	University
Yen Lin-shan	阎林山	Observatory
Ying Chih-hsiang	应志祥	Observatory

OCTOBER 18-19, NANKING; HISTORIANS OF SCIENCE

Chang P'ei-yü	张培瑜	Leading Member, Research Group in Ancient Astronomy, Purple Mountain Observatory
Chang Yü-che	张钰哲	Director, Observatory
Ch'e Yi-hsiung	车一雄	Research Group
Wang Te-ch'ang	王德昌	Research Group
Hsü Chen-t'ao	徐振韬	Research Group
Li T'ien-tz'u	李天赐	Research Group

APPENDIX B

Summary Itinerary

September 27-28   Delegation gathering in Tokyo, Keio Plaza Hotel

September 28   Tokyo Observatory, Mitaka

September 29   Noon departure on *Air France* via Fukawara for
                   Peking; late afternoon arrival at Peking Hotel,
                   old wing

September 30

Morning   Sha-ho Station of Peking Observatory, in Sino-
                   Vietnamese People's Friendship Commune about
                   45-m drive northwest of Peking; primary solar and
                   time-keeping work

Afternoon   Museum of Chinese History

Evening   Dinner in the Great Hall of the People, hosted
                   by Chairman Hua Kuo-feng for several thousand
                   Chinese and foreign guests

October 1

Morning   Summer Palace, for National Day celebration

Afternoon   Temple of Heaven, for National Day celebration

Evening   Tien An-men Square, reviewing stand, for folk
                   dancing and fireworks

October 2

Morning   Great Wall

Afternoon   Ming Tombs, Sports Palace

October 3

Morning   Western Hills, Fragrant Hill Park

Afternoon   Forbidden City

October 4

Morning   Hsing-lung Station (stellar astronomy) of Peking
                        Observatory, 960-m elevation in Yen Sen Mountains,
                        about 4 h drive northeast of Peking

Afternoon   Half of group remains at Hsing-lung overnight;
                        remainder return to Peking

October 5

Morning   Mi-yun Radio Branch of Peking Observatory, about
                        3½h drive north of Peking

Afternoon   Return to Peking

October 6

Morning   Brief visit to Chinese Academy of Sciences Library

                        Chinese astronomy lectures in third-floor lecture
                        hall of the Institute of Biophysics

Afternoon   Continue lectures

Evening   Several small discussion groups with astronomers

October 7

All day   Continue lectures at Chinese Academy of Sciences

Evening   Reception at U.S. Liaison Office

                        Banquet at Peking Roast Duck Restaurant

October 8

Morning	Peking University
Afternoon	Large-group joint discussions with Chinese astronomers at Peking Hotel
Evening	Several small discussion groups with astronomers

October 9

Morning	Tien An-men Square: Mao Tse-tung's Mausoleum
	Continue large discussion groups at Peking Hotel
Afternoon	Shopping, sight-seeing around Peking
Evening	Farewell (to Peking) banquet at Szechwan Restaurant

October 10

Morning	Flight from Peking to Kunming; check in at Kunming Hotel
Afternoon	Yünnan Observatory, about 9-km north of Kunming
Evening	Special movies at hotel (monkey traditional tale, gymnasts)

October 11

All day	Lectures by Chinese and American astronomers at hotel
Evening	Vaudeville show in Kunming municipal theater

October 12

All day	Stone Forest, about 4 h drive northwest of Kunming; old temples on cliff south of lake
Evening	Special movies at hotel (Pandas, Wildlife of Yünnan, Snake Island)

October 13

Morning   Scientific group discussions at hotel

Noon   Flight from Kunming to Shanghai via Changsha;
                   check in at Chin-chiang Hotel

Evening   Sight-seeing around Bund

October 14

All day   Shanghai Observatory (mainly time-keeping), near
                   center of city; lectures by Chinese and American
                   astronomers

                   Part of the group visits the Huang Tu Commune and
                   the Shanghai Optical Instruments Factory

Evening   Banquet at hotel

October 15

Morning   Lectures at Shanghai Observatory

Afternoon   Train from Shanghai to Nanking

Evening   Check in at Nanking Hotel

October 16

Morning   Purple Mountain Observatory (stellar, solar
                   system, history)

Afternoon   Sun Yat-sen Memorial, Pagoda, Yangtze River Bridge

Evening   Banquet at Nanking Hotel guest house

October 17

Morning   Nanking University and Astronomy Department

Afternoon   Nanking Astronomical Instruments Factory

Evening   Acrobatics performance

October 18

All day   Lectures by Chinese and American astronomers at
                 Purple Mountain Observatory

Evening   Discussions at hotel about astronomy education
                 with astronomy faculty of Nanking University

October 19

All day   Continue astronomy lectures at Purple Mountain
                 Observatory

Evening   Discussions at hotel with personnel of Nanking
                 Astronomical Instruments Factory concerning large
                 telescopes in general and the new Chinese 2-m
                 telescope now under construction

October 20

Morning   Discussions at hotel with stellar astronomy group

Afternoon   Train from Nanking to Shanghai

Evening   Flight from Shanghai to Canton; check in at Tung
                 Fang Hotel (new wing)

October 21

Morning   Flight from Canton to Kweilin; check in at Li
                 Chang Hotel

Afternoon   Visit two caverns

Evening   Banquet at hotel

October 22

Morning   Drive to riverboat dock through unreal karst
                 mountains

Afternoon   Boat ride down river through even less real
                 mountains

Evening   Banquet at hotel in honor of our guides

October 23

Morning   Kweilin City sight-seeing

Afternoon   Flight from Kweilin to Canton; check in at Li
        Chang Hotel (old wing)

Evening   Farewell banquet at White House on the Lake
        Restaurant (seats 3,000 at a time !)

October 24

Morning   Depart Canton by train to Hong Kong; end of
        official trip

SEPARATE AGENDA ITEMS FOR NATHAN SIVIN

Peking

October 6

All day   Meetings with historians of science for
        lectures and discussions

October 7

Morning   Meetings with historians of science for
        informal discussions

Afternoon   Visit to rare book collection, Peking Library

October 8

Afternoon   Meetings with historians of science for
        informal discussions

Evening   Dinner with four members of Research Institute
        for the History of Natural Sciences, Chinese
        Academy of Social Sciences .

October 9

Morning   Meetings with historians of science for
        informal discussions

Afternoon   Visit to bookstores

Shanghai

October 14

Afternoon   Meetings with historians of science

October 15

Morning   Visit to bookstores

Nanking

October 18

All day   Meetings with historians of science for
                             lectures and discussions

October 19

Morning   Visit to Nanking Museum

Afternoon   Meetings with historians of science for
                             informal discussions

APPENDIX C: LECTURES

LECTURES PRESENTED BY THE CHINESE

Peking

Chinese Academy of Sciences (CAS)

"Abnormal Neutron Stars," Fang Li-chih, Chinese University of Science
    and Technology; Ch'ü Ch'in-yueh, Wang Chen-ju, Nanking University;
    Lu T'an, Telecommunication Instruments Factory, Nanking; Lo L'iao-fu,
    Inner Mongolian University.
"The Galactic Shock Waves with Self-Gravitational Force of Interstellar
    Gas," Hu Wen-jui, Institute of Mechanics, CAS; Ao Chao, Institute
    of Computation, CAS.
"Jeans Instability of the Gas Disk with Two Components," Chou
    Tseng-lung, Ch'en Chien-sheng, Peking Observatory, CAS.
"The Phase Shift Between Star Arm and Gas Arm in Spiral Galaxies," Liu
    Yung-chen, Fang Li-chih, Chinese University of Science and Technol-
    ogy.
"On the Expansion of the Spiral Galaxies," Li Ch'i-pin, Peking
    Observatory, CAS.
"On the Redshift Distribution and the Luminosity Function of Extra-
    galactic Objects," Sun K'ai, Astronomical Section, Department of
    Geophysics, Peking University.
"Statistical Analysis of QSO's with Radio Components," Chou Yu-yuan,
    Fang Li-chih, Cheng Fu-chen, Chu Yao-ch'üan, Cheng Fu-hua, Chinese
    University of Science and Technology.
"Observations of the Gravitational Effects During the Annular Eclipse
    of 29 April 1976, Part I: Results Obtained with the Gravimeter and
    the Inclination Indicator," Wang Liu-ch'uan, T'ien Ching-fa, Liu
    Yi-fen, T'ang Shao-lin, Chao Chi-shu, Chin Yung-hsien, Institute
    of Physics, CAS; Liu Yi-cheng, Institute of Mathematics, CAS; Chang
    Chien-chao, Earthquake Team of Sinkiang.
"Some Work on Solar Physics at Peking Observatory," Shih Chung-hsien,
    Peking Observatory, CAS.
"On the Mechanism of Type-I Bursts," Liu Hsü-chao, Peking Observatory.
    CAS.

"Evolutionary Relation Between the Sources of Type-I Bursts of Meter
    Wavelength and the Flare Activities," Yin Ch'i-feng, Lo Shao-kuang,
    Astronomical Section, Department of Geophysics, Peking University.
"On the Dynamics of the Formation of the Planetary System," Li Ch'i-pin,
    Peking Observatory.
"Some Research on the Gauge Theories of Gravitation," Kuo Han-ying,
    Ch'en Shih, Ho Tso-hsiu, Institute of High Energy Physics, CAS;
    An Ying, Li Ken-tao, Chang Li-ning, Institute of Mathematics, CAS;
    Chang Yuan-chung, Wu Yung-shih, Institute of Physics, CAS; Chou
    Chen-lung, Huang P'eng, Peking Observatory, CAS.
"Typical Space-Time Theory and Its Effects on Cosmology," Lu Ch'i-keng,
    Chang Li-ning, Institute of Mathematics, CAS; Kuo Han-ying, Insti-
    tute of High Energy Physics, CAS; Chou Tsen-long, Ch'en Chien-sheng,
    Huang P'eng, Peking Observatory, CAS.
"On the Solution of Light Curve of Partially Eclipsing Binary System,"
    K. H. Look, Institute of Mathematics, CAS; Cheng Chien-sheng,
    Chou Chen-lung, Peking Observatory, CAS.
"Spectrographic Observations of Nova Cygni 1975," Liu Tsung-li *et al.*,
    Peking Observatory, CAS.
"Some Results of Photoelectric Observations of the Eclipsing Binary
    CG Vir," Shen Liang-tsao, Che Ti-shen, Lin Po-sen, Chiang Chao-chi,
    Peking Observatory, CAS.
"Photoelectric Photometry During the 1975 Eclipse of AZ Cas," Li Feng,
    Chiang Shih-yang, Kuo Chi-ho, Hsu Hui-fang, Che Ti-shen, Shen
    Liang-tsao *et al.*, Peking Observatory, CAS.

Peking University

"The Distribution of Mass Density in a Disk Galaxy of Finite Thickness,"
    P'eng Ch'iu-ho.
"The Stable Confinement of Fast Electrons in the Region of Type I
    Bursts of Meter Wavelength," Yin Ch'i-feng, Ma Erh.
"A Possible Design for a Millimeter-Wave Solar Radioheliograph,"
    An Ching-chu, Wang Shou-kuan, Ch'iao Ku-chüm.
"The Formation of the Basic Solar Magnetic Field and Sunspot Activity,"
    Chou Tao-ch'i.
"The Wilson Effect in Sunspots," Yang Hai-shou.

Kunming

Yünnan Observatory

"To Investigate the Stability of the Sunspot Period by Means of the
    Periodic Analysis of the Aurora and Earthquake Records in Ancient
    China," Lo Pao-jung, Li Wei-pao, Yünnan Observatory, CAS.
"The Asymptotic Solution of the Poisson's Equation of the Screw-Shaped
    Disturbance of the Density of the Spiral Galaxy with Limited
    Thickness," P'eng Ch'iu-ho, Department of Geophysics, Peking
    University; Huang Yin-liang, Yünnan Observatory, CAS.

"The Spiral Sunspots in Solar Active Region in October 1972," Ting Yu-chi, Yünnan Observatory, CAS.

"Ancient Records of Sunspots in China 43 BC-AD 1638," Chang Chu-wen on behalf of the Group of Arrangement of the Chinese Ancient Sunspot Records, Yünnan Observatory, CAS.

Shanghai

Shanghai Observatory

"A Correlation Analysis of X-ray Luminosity and Cluster Velocity Dispersion," Wan Lai, Shanghai Observatory, CAS.

"On the Stability of the CIO System, Chao Ming, Lu Chu-ying, Shanghai Observatory, CAS.

"The Distribution in Space of Clusters of Galaxies and of Radio Sources," Chao Chün-liang, Shanghai Observatory, CAS.

"Some Features About the Local Non-polar Terms of Latitude During the Years 1949-73," Li Cheng-hsin, Shanghai Observatory, CAS.

"A Stable Rotational Measure and Other Properties of Galaxies," Tai Wen-sai.

"Stellar Motions in Selected Clusters and Associated Galaxies," Ch'ien Po-ch'en, Shanghai Observatory.

Nanking

Purple Mountain Observatory

"The Photoelectric Observation of Occulation of SAO 158687 by Uranian Ring," Planetary Laboratory, Purple Mountain Observatory, CAS; Stellar Division, Peking Observatory, CAS.

"The Theory of Radiative Transfer of Uranian Ring," Ch'en Tao-han *et al.*, Planetary Laboratory, Purple Mountain Observatory, CAS.

"The Mathematical Statistical Detection of Uranian Ring-Signals from the Light Curve," Ch'en Tao-han, Wu Yueh-chen, Purple Mountain Observatory, CAS; Chiang Shih-yang, Peking Observatory, CAS; Hsu Chung-che, Cheng Chien-shen, Peking University; Chang To-chun, Institute of Geography, CAS.

"Porcelain Mirror," Yang Shih-chieh, Purple Mountain Observatory, CAS.

"Magnetic Field in a Coronal Condensation Region," Ts'ao T'ien-chun, Purple Mountain Observatory, CAS.

"On Gaseous Solitons of a Two-Component Disk and an Explanation to the Titius-Bode Rule," Li Hsiao-ch'ing, Purple Mountain Observatory, CAS; P'eng Ch'iu-ho, Department of Geophysics, Peking University; Su Hung-chün, Purple Mountain Observatory, CAS; Huang K'o-liang, Huang Chieh-hao, Department of Astronomy, Nanking University.

"The Theory of Gravitational Instability of Polytropic Gaseous Bodies," Hiroshi Kimura and Liu Ts'ai-p'in, Purple Mountain Observatory, CAS.

"Observations of Flare Stars in the ρ Oph Cloud and the Taurus Cloud Regions," Third Group of the Stellar Physics Division, Purple

Mountain Observatory, CAS; Huang Cheng-chun, First Group of the Stellar Physics Division, Peking Observatory.

"The General Properties of Hot Ultrashort Period Cepheids," Ch'u Yu-hua, Purple Mountain Observatory, CAS.

"Historic Records about the Tien-Kuang Guest Star of AD 1054," Wang Te-chang, Purple Mountain Observatory, CAS.

"X Persei and the X-ray Source 3UO352+30," Hang Heng-jung, Mo Ching-erh, Purple Mountain Observatory, CAS.

"The Influence of Electron-Positron Pairs on the Accretion Disk and the Spectrum of CYG-X-1," Chang Chia-lü, Fang Li-chih, Chinese University of Science and Technology.

"The Anti-matter Model of Quasars and Active Galactic Nuclei," Lu T'an, Telecommunication Instruments Factory of Nanking; Lo-Liao-fu, Inner Mongolian University; Ch'ü Ch'in-yueh, Wang Chen-ju, Department of Astronomy, Nanking University.

"The Separation Velocity of the Radio Components of QSO's," Yu Chun-han, Cheng Fu-chen, Fang Li-chih, Chinese University of Science and Technology.

"Properties in Relativistic Electron Gas," Chang Ho-ch'i.

"The Orbital Evolution and History of Halley's Comet," Chang Yü-che.

"The Lunar Eclipses and The Calendar Under the Reign of Wu Ting in the Yin Dynasty," Chang P'ei-yü.

"The Han Folding Gnomon," Ch'e Yi-hsiung, Research Group in Ancient Astronomy, Purple Mountain Observatory.

Nanking University

"The Statistical Analysis of Pulsars and JP 1953," Ch'ü Ch'in-yueh, Wang Chen-ju, Department of Astronomy, Nanking University; Lu T'an, Telecommunication Instruments Factory of Nanking; Lo Liao-fu, Inner Mongolian University.

"The Model of the Chromosphere Heating by Shock Waves," Yao Pan, Department of Astronomy, University of Nanking (read by Hsu Ao-ao).

Nanking Instrument Factory

"China-Made Photographic Zenith Tube," Ku Chen-lei, Nanking Astronomical Instruments Factory.

"Photoelectric Astrolabe Type II," Hu Ning-sheng, Photoelectric Research Group.

"400mm Horizontal Solar Telescope and Multiple-Bands Spectrograph," Li T'ing, Nanking Instrument Factory.

"Automatic Design of Corrector Systems for Cassegrain Telescopes," Su Ting-chiang, Yü Hsin-mu, Nanking Astronomical Instrument Factory, CAS; Wang Lan-chuan, Yeh Chih-feng, Shanghai Observatory, CAS.

"Mechanical Design of Two-Meter Telescope," Kuo Nai-shu, Nanking Astronomical Instruments Factory.

LECTURES PRESENTED BY THE DELEGATION

Peking

"Infrared Astronomy," Charles Townes.
"Microwave Receivers and Astronomy," Charles Townes.
"Mass Loss From Late Type Giants and Supergiants," Leo Goldberg.
"Solar Physics at Kitt Peak National Observatory," Leo Goldberg.
"Compact Radio Sources," D. S. Heeschen.
"The Very Large Array," D. S. Heeschen.
"Radio Redshifts," D. S. Heeschen.
"Extended Radio Sources," D. S. Heeschen.
"The Texas 360-MHz Array and Radio Source Catalog," Harlan J. Smith.
"Spectra of QSOs," E. Margaret Burbidge.
"Velocity-Distant Relations and the Value of q_0," E. Margaret Burbidge.
"Stellar Evolution and the Galactic Evolution," Allan R. Sandage, E.
    Margaret Burbidge, and Martin Schwarzschild.
"The Problem of the Redshifts," Allan R. Sandage and E. Margaret Burbidge.
"Optical Redshifts and the Hubble Expansion," Allan R. Sandage.
"Interpretation of Quasar Redshifts as Cosmological," Allan R. Sandage.

Yünnan Observatory

"Astronomical Observing Site Selection," Victor Blanco.
"Recent Developments in Solar Physics in the U.S., Including Work on
    Small- and Large-Scale Magnetic Fields and Global Oscillations,"
    Leo Goldberg.
"Late Stages of Stellar Evolution, Neutron Stars, and Black Holes,"
    Martin Schwarzschild.
"Spectra of QSOs - Redshifts and Physical Conditions," E. Margaret
    Burbidge.
"Twenty-one cm Observations in Galaxies," David S. Heeschen.
"Modern Detectors and Instrumentation for Faint Objects," George H.
    Herbig.
"Redshifts, Cosmology, and the Hubble Expansion," Allan R. Sandage.

Shanghai Observatory

"Radio Astrometry," D. S. Heeschen.
"Lunar and Satellite Laser Techniques for Geophysics and Timekeeping,"
    Harlan J. Smith.

Purple Mountain Observatory

"Solar Magnetic Fields and Global Oscillations," Leo Goldberg.
"The Circumstellar Shell Around Alpha Orionis," Leo Goldberg.
"Red Giant Stars in the Galactic Center and the Magellanic Clouds,"
    Victor Blanco.

"New Developments in the Study of Pre-Main Sequence Stellar Evolution,"
   George Herbig.
"U.S. Astronomy Education," Harlan J. Smith and Martin Schwarzschild.
"Solar System Space Astronomy," Harlan J. Smith.
"Mass Loss from Stars," Martin Schwarzschild.
"Active Nucleii of Galaxies," E. Margaret Burbidge.
"Radio Observations of Nucleii of Galaxies," David S. Heeschen.

LECTURES PRESENTED IN MEETINGS OF NATHAN SIVIN WITH HISTORIANS OF
SCIENCE

Lectures Presented by the Chinese

Peking

"Archeological Discoveries in the History of Astronomy Made Since the
   Beginning of the Cultural Revolution," Hsia Nai, Director, Research
   Institute of Archeology, Chinese Academy of Social Sciences.
"Special Characteristics of Ancient Chinese Mathematical Astronomy,"
   Yen Tun-chieh, Research Institute of the History of Natural Sciences,
   Chinese Academy of Social Sciences.
"Knowledge of Planets and Comets Reflected in the Silk MSS Found in the
   Han Tomb at Ma-wang-tui," Hsi Tse-tsung, Research Institute of the
   History of Natural Sciences, Chinese Academy of Social Sciences.

Kunming

"Historical Sunspots and the Periodicity of Solar Activity,"
   Chang Chu-wen, Kunming Observatory.
"Stability of Sunspot Periods by Periodic Analysis of Aurora and
   Earthquake Records," Lo Pao-jung, Li Wei-pao.

Shanghai

"On the Accuracy of Chinese Clepsydras of the Ch'in and Han Periods,"
   Ch'üan Ho-chün, Yen Lin-shan, Shanghai Observatory.
"On the Hundred-Mark Time-Reckoning System Indigenous to China,"
   Yen Lin-shan, Ch'üan Ho-chün, Shanghai Observatory.

Nanking

"The Orbital Evolution and History of Halley's Comet," Chang Yü-che,
   Director, Purple Mountain Observatory.
"The Lunar Eclipses and the Calendar Under the Reign of Wu-ting in the
   Yin Dynasty," Chang P'ei-yü, Research Group in Ancient Astronomy,
   Purple Mountain Observatory.
"The Han Folding Gnomon," Ch'e I-hsiung, Research Group in Ancient
   Astronomy, Purple Mountain Observatory.
"Historic Records of the T'ien-kuan Guest Star of AD 1054," Wang
   Te-ch'ang, Research Group in Ancient Astronomy, Purple Mountain
   Observatory.

Lectures Presented by Sivin

"Chinese Responses to the Introduction of Western Science in the
Seventeenth and Eighteenth Centuries."
"Research Under Way on the History of Chinese Mathematics and
Astronomy in Western Europe, the United States, and Japan."
"Research Under Way on the History of Other Fields of Chinese Science
in Western Europe, the United States, and Japan."

APPENDIX D: COMMENTS ON THE SELECTION OF
AN OBSERVING SITE FOR OPTICAL ASTRONOMY

In the selection of an observing site, many aspects must be considered.
Some of these, such as accessibility and availability of power and
water, are of a logistic or economic nature. Other considerations
have to do with the type of observations to be made and the quality of
the atmosphere. In general, for nighttime observations the following
characteristics are desirable: absence of clouds, darkness of the night
sky, transparency, good seeing, dryness, and absence of manmade light or
other radiation interference. These comments summarize the desirable
atmospheric characteristics for optical work.

CLOUD COVERAGE

Climatological records and satellite earth-photography can give us
valuable information about cloudiness; however, nighttime conditions,
especially in remote areas, are frequently not known, and the available
information must be supplemented with weather observations. These
should be carried out for at least 2 years. In regard to cloudiness,
the observations should determine among other things the fraction of
the visible sky covered by clouds. For convenience we say that if the
sky has less than one-eighth of cloud coverage during at least half of
the night, the night is "observational." If the cloud coverage is less
than one-eighth throughout the entire night, the night is
"photometric." At a good site the number of observational nights is
greater than about 270 per year and at least 60 percent of these are
photometric.

SKY DARKNESS

The brightness of the night sky determines the irreducible noise level
affecting any astronomical observation. The brightness can be measured with common photometric techniques if the known brightness of a
given star is used as reference. Short-focus telescopes and photometer diaphragms that cover up to one square degree of the sky are
suitable for this purpose. In a good site away from the Milky Way and
the zodiacal light, a B-magnitude approximately equal to or fainter

106

than 22.5 and a V-magnitude equal to or fainter than 21.5, within 30°
of the zenith, are not uncommon. The absence of appreciable city
lights is important. As emphasized by M. F. Walker (PASP 82, 672, 1970,
and PASP 85, 508, 1973), observatories should be located away from
large population centers and from towns that are likely to grow
markedly in population. If some distant city lights are unavoidable,
their effect on astronomical observations can be minimized by proper
choice of outdoor lighting. For example, lighting fixtures that radi-
ate upward as well as in the blue and ultraviolet spectral regions
should be avoided.

TRANSPARENCY

At a good site the atmosphere should be very transparent. A good daytime indicator of transparency is the deep blue color caused by
Rayleigh scattering. This so-called "coronal quality" sky is commonly
seen at good sites. It may be tested for by observing the amount of
scattered light close to the Sun's limb if the Sun is covered by one's
thumb. Photometric absorption coefficient should also be reasonably
small. In the UBV photometric system, acceptable upper limits are:
$\kappa_V = 0.17$, $\kappa_B = 0.028$, $\kappa_{U-B} = 0.32$.

SEEING

Astronomical "seeing" has not been well-defined. Different methods
used to observe it may measure various effects which are necessarily
related to each other (J. Stock and G. Keller, *Telescopes*, G. P. Kuiper
and B. M. Middlehurst, eds., Chicago: University of Chicago Press,
1960, p. 138; also A. B. Meinel, *ibid.*, p. 154). Although the effects
produced by random changes in the direction of star beams are generally
called *seeing*, and random variations in stellar image intensity
scintillation, the term "seeing" may be used to talk about both phenomena. With relatively small telescopes a stellar image may be observed
to show random displacements as well as scintillation. These are
caused by atmospheric inhomogeneities that may originate locally or in
the upper atmosphere. With large telescopic apertures such displacements and intensity variations may not be observed. Instead, the
image size is often seen to be larger than as seen with small telescopes. This is caused by the fact that atmospheric inhomogeneities
are relatively small in size. With telescope apertures larger than
about 1 m, the effects of a sufficiently large number of such inhomo-
geneity cells are integrated. The knife-edge or Foucault test in
relatively large reflectors shows clearly the effects and the average
size of such inhomogeneities. In such tests, local seeing (sometimes
called dome-seeing) in general produces slow-moving shadow patterns
that may be easily seen to vary in shape as they cross the image of the
primary mirror. Upper atmosphere (also called *meteorological*) seeing
shows swift moving and relatively small shadow patterns.

Atmospheric turbulence and thermal differences in the airflow have been found to be correlated with seeing. In site testing, tower-mounted high-speed thermal detectors and anemometers are useful for measuring the quality of the atmosphere near the ground. The ideal conditions that one looks for are very small temperature variations (say under 1°C over about 5-min intervals) and laminar wind flow. The net effects of local and meteorological seeing may be determined in a number of different ways with telescopic apertures of about 15-30 cm. These include A. Danjon's method (*C. R. Acad. Sci. Paris* 183:1032, 1926) based on the study of diffraction patterns, image motion studies in which the motion component perpendicular to the diurnal motion is measured, and the double beam method, in which the relative motion and image forms produced by two separate apertures are studied. A simple method that is now often used consists of observing the nature of polar star trails as described by E. A. Harlan and M. F. Walker (ASP Publ. No. 77, p. 246, 1965). Small telescope site-testing methods must be calibrated with the aid of relatively large reflectors. For example, in the Harlan-Walker star-trail method a correlation was established between star-trail characterisitcs and image size as seen at the coudé focus of Lick Observatory's 3-m telescope. The image size produced by a large telescope may be conveniently measured in a well-guided photograph in order to avoid photographic effects that affect image size, the diameter of relatively faint stellar images is used. In a good site this so called *seeing image diameter* is on the average less than 2 arc sec, and averages on the order of 1 arc sec may be found at the best sites.

Local seeing is adversely affected by a chimney effect within the dome. Dome design and construction should aim at equalization of the temperature of the telescope's objective with that of the nighttime outdoor air. Double-shell domes and fans or ventilators are very helpful in achieving this equalization. Local seeing depends on orographic wind effects. The local terrain can cause undesirable eddies. The ideal mountain site will affect laminar wind flow the least.

Meteorological or nonlocal seeing is correlated with turbulent wind flow, and the latter is prevalent when the lapse rate is adiabatic or super-adiabatic in the lower atmosphere. In the better sites known so far, semipermanent ground inversion layers are found that may have a depth of 1 km or more. The lack of atmospheric turbulence associated with such layers is correlated with a very low frequency of towering cumulus clouds and of thunderstorms, as well as with the existence of haze in the lower atmospheric layer. Experience shows that a good mountain observing site has an elevation at least 500 m higher than the top of such low-level inversion layers. A meteorological condition that causes inversion layers, or at least nonturbulent lapse rates, is subsidence in high pressure areas, as emphasized by B. McInnes, M. Hartley and T. T. Gough (*The Observatory* 94:14, 1974).

DRYNESS

Observations in the infrared spectral regions require relatively dry atmospheric conditions. At better sites the amount of precipitable atmospheric water vapor is about 3 mm or less during a major fraction of the year. Atmospheric water vapor may be easily determined with a device that measures the strength of molecular water vapor bands in solar observations (O. L. Hansen and L. Caimanque, ASP Publ. No. 87, p. 935, 1975). Dryness is also important for preventing undesirable condensation which may affect equipment operation.

MISCELLANEOUS CONSIDERATIONS

It is desirable to have reasonably small diurnal temperature variations, although this condition is generally incompatible with the existence of clear skies. Appropriate dome-design can minimize considerably the effects of diurnal temperature variations. More important is the temperature variation of the air in the period between the evening and morning astronomical twilights, since such variations will affect the telescope when the dome is normally open. Such temperature ranges may average less than 5°C at a good site. Low average wind speeds are also desirable, especially if the local terrain can cause turbulent eddies.

Absence of nearby city lights has been mentioned already. A related desirable feature, and one within human control, is the absence of high-power radio or television transmitters in the observatory's vicinity. Modern instrumentation is unusually susceptible to such radiations. Radio communication with the observatory when necessary can be properly handled with highly directional low-power (2 W) microwave beams.

DATE DUE